50 golden gardenin

by Roger Aylett

Published in the UK for Aylett Nurseries Limited,
North Orbital Road, London Colney, St Albans, Hertfordshire AL2 1DH
by Burall Floraprint Limited, Wisbech, Cambridgeshire PE13 2TH

First Edition 2005

A CIP catalogue record for this book is available from the British Library

ISBN 0-903001-73-X

Text by Roger Aylett

Designed and Typeset by Louise Asquith, Asquith Design

Illustrations by Jim Housego

Front Cover photograph by Hazel Aylett

Printed in Great Britain

'My interest in dahlias started at a very early age when as a young boy I grew them in the back garden at home and gradually horticulture became a serious hobby and later a career ... the seeds were sown!'

where it all began...1955

50 Golden Gardening Years
is dedicated to the
memory of my dear parents
and to my daughters
Julie and Sally
and their children.

foreword ...

Our generation has experienced more changes than any of our forbearers, from horsepower to supersonic flight. Family businesses have remained Britain's lifeblood throughout this period, and it is interesting to read how the ideas, energies, investments and dogged determination of one person can generate so many benefits for a community.

All nurserymen and gardeners will enjoy reading Roger's autobiography as much as I, his contemporary, have done. What, I wonder, will today's college students make of an hours practical work before breakfast and routine Saturday lectures? Will they build businesses employing 160 people?

Those of us who work the flower show circuit are in awe of The Royal Horticultural Society's Williams Medal, given each year for the finest exhibit of one plant species. Win this and you've really made it: Aylett Nurseries have won it three times – what more can be said?

The Shrewsbury Floral Fete was – in the days of Percy Thrower – the tops. I slept overnight alongside the river as Roger recalls he and Hazel did – no comfy caravan for me however, just the not-too-watertight canvas tilt on the 5-ton Bedford lorry! How would we have managed in the 50's and 60's without all that Army surplus telephone wire and why was it surplus?

Thank you Roger for chronicling the remarkable Aylett achievements over the past 50 years, reminding me of happy days.

Peter Seabrook V.M.H.

chapter **1**

...we will not move out of london whatever happens

9

I was born in Blackheath, South London in 1934. My parents lived in Sydenham and my Father commuted to London each day. At that time there was a lot of talk about war and Father's chairman said, 'We will not move out of London, whatever happens.'

But in September 1939, when war was declared on Germany and the bombing started on London, Father was asked by his company to go to St. Albans to look at a factory which could be converted, ensuring that production of the famous Rose's Lime Juice could continue. He was successful and the site in Grosvenor Road was purchased. So a month later we moved to St. Albans and for the first three weeks we lived in a pub called 'The Peahen'. It was while we were staying there that my Mother made blackout curtains, to be put up at the windows of our new house in Marshalls Drive. To one side of the house was a field, about an acre in size. There was a little wood in the centre, in which I played very happily. I began to take an interest in the things that were around me, such as Mother's chickens, which she reared for their eggs, as these were extremely scarce during the war.

Mother accompanied my Father to London one day and while he was dealing with business, she decided to take a trip to Harrods. I think it demonstrates her great character when I tell you that her main purchase from the pet department was some day-old Bantam hens. It became evident, as these chicks matured, that one of them was of the male gender. He thought this was a good life and made friends with the Rhode Island Reds.

...and for the first three weeks we lived in a pub called 'The Peahen'

We soon had a clutch of chicks and as they developed you could see that they were a cross between the Bantam and the Rhode Island Red. This was to be my first lesson in hybridising. We also had some Khaki Campbell ducks and these produced lovely blue eggs which I enjoyed eating. They made such an impression on my young life that many years later, when I had children of my own, I bought some two dozen day old chicks, but that's another story.

On one occasion my Father came home with a goat, with the idea of producing goat's milk. But alas, he failed to purchase a pair, so I'm not quite sure how he thought that was going to come about. It all went sadly wrong and the only task the goat was interested in was helping my Grandmother across the field by butting her. She was not in the least amused and set about the goat with her umbrella. Needless to say, the goat did not last long and soon had to find a new home, but while he was with us he caused quite a stir.

12   *...not in the least amused and set about the goat with her umbrella*

During this difficult time, my parents grew many vegetables to help supplement their wartime rations. I found this fascinating and they encouraged me to have my own vegetable plot, where I raised carrots, beetroot, runner beans and potatoes. As I grew older, I added sweet peas, dahlias and chrysanthemums to my cropping programme.

My Father bought a greenhouse for us. It measured 25ft by 10ft and was heated by a little coal-fired boiler situated on the outside wall. The fire heated the water, which circulated in a pipe under the staging and kept the night temperature somewhere between 45°F and 50°F. Each night I made up the fire and Mother helped me with this chore in the morning, as I had to go to school.

Marshall Drive greenhouse

I grew many seed crops and pricked them out and grew the little seedlings on. During May I sold my boxes of plants to friends and neighbours for a modest price. I used the money to buy more seeds and although I daresay Father subsidised me to a great extent, I began to fund my little project myself. I bunched the cut flowers that I grew and began to sell them to greengrocers in St. Albans, except for odd weekends when there was a flower show in which I wanted to compete. Harpenden Horticultural Society's flower show took place in the Public Hall in September and the St. Albans Horticultural Society's flower show was held in Culver Hall, which sadly, like the society itself, no longer exists.

When I was about 15, I took my produce to Culver Hall for the annual show and entered the classes for dahlias, chrysanthemums, runner beans and tomatoes. As I arranged my plate of tomatoes, I noticed that the other exhibitors had picked their fruit with the calyx attached. I had removed my tomatoes' calyxes and realized that this was a mistake. I sped home on my bicycle and picked a fresh supply of calyxes. I found a pot of glue and went back to the hall where I made the necessary adjustments. Unfortunately the glue failed to dry before the judges arrived and my exhibit was consequently disqualified. This was my first lesson in not trying to pull the wool over judge's eyes.

. . . unfortunately the glue failed to dry before the judges arrived

Through the remainder of my school years I won many prizes for my exhibits, and gradually became more confident. I attended St. Georges School in Harpenden, where it has to be said I was not the academic type, although I did excel in Maths, Biology and Geography. I thoroughly enjoyed my time there, and as time progressed it began to dawn on me that my increasingly engrossing hobby could easily be turned into a career. So I set about finding myself some suitable training in this field.

... set about finding myself suitable training

To gain my horticultural qualification, I wanted to go to Oaklands, which at that time was one of the leading horticultural institutes in England and just happened to be based in Hatfield Road, St. Albans. It's known today as Oaklands College, and it offers a wide spectrum of educational courses on various campuses throughout Hertfordshire.

Before I could start on my chosen course, I had to complete a year's practical training. This stipulation applied to all agricultural and horticultural courses and was aimed at ensuring that all students had at least a smattering of knowledge of the subject and, just as importantly, that they were made of the 'right stuff'. The assumption was, that anyone prepared to do a year's work for no payment and to complete their basic tasks diligently and obediently, must be serious about their course and likely to complete it successfully. Although young people today may find it hard to believe, students at that time expected no wages – they were only too pleased to be in a position to absorb valuable knowledge from their seniors and to secure their place on a much sought-after course.

I spent my first few months at Childwickbury, a wonderful estate owned by the late Solly Joel, who was a very talented racehorse breeder. Mr. Daffern was the head gardener and he helped me with my initiation into the working horticultural world. I helped out in the walled garden, where there were three lean-to greenhouses. On the back walls, grapes, nectarines and peaches were trained and the beds below them were given over to a flourishing range of seed crops such as melons, early lettuce, carrots and spring onions.

Orchids had a house all to themselves and I spent a good deal of time sponging the leaves with an insecticide to control scale insects. These are particularly prevalent in orchids and they're the very devil to remove. My favourites were the cymbidium orchids, with their long stems carrying the many flowers. These were used to decorate the house when Mr. Joel was entertaining, which he did in a grand style. Mr. Daffern and his assistant, Charlie Ellis, spent endless hours decorating the house ready for the weekend parties and although I always hoped I'd be invited to help with these preparations, I was invariably disappointed.

cymbidium orchids

After my time at Childwickbury, I moved on to a commercial concern called Puckeridge Park. It was part of an estate owned by the accomplished musician, Sir Francis Cassel and I remember that while I was employed there, some plants were grown to decorate the Albert Hall for one of his concerts. The estate was just outside Luton and if she wasn't using it herself, I would borrow my Mother's car and drive myself there. Mr. Roy Brockway, who was a great help in my training, managed the commercial enterprise. He encouraged all his younger members of staff, patiently taking the time to explain the jobs, such as planting early flower chrysanthemums which were grown as a cut flower crop and sold in various markets such as Covent Garden. I also spent time trimming cucumbers and removing the male flowers, as this was before all-female cucumbers had been bred. I can never forget the considerable length of time I spent potting

young Primula obconica. This work was done by hand, as potting machines were unheard of in those days. I must have spent five weeks doing nothing else and when I had nearly finished the crop I developed a very uncomfortable primula rash. My time at Puckeridge Park stood me in good stead for my arrival at Oaklands, in September 1953, to start my formal horticultural training.

... my arrival at oaklands september 1953

The first year of the course was devoted to commercial horticulture. The day began at 7 o'clock in the morning, when an hour's practical work was completed before breakfast. As I was a dayboy and not a boarder, I was excused this duty. I have to admit that this caused a little resentment amongst my fellow students, who thought the arrangement was grossly unfair.

Breakfast was followed by lectures on a wide spectrum of horticultural subjects and every Saturday morning we had to complete an identification of thirty different specie of plants. Each student was allocated a small plot of land, which had to be cultivated and cropped by themselves in their spare time. This formed part of the overall course and could mean the difference between pass and failure, so it was only a very short-sighted student who didn't keep their plot up to standard.

The afternoon timetable concentrated on commercial crops and alternate weeks would be spent in the market garden and the glasshouse department. In the market garden, gladioli were planted between lettuces in Dutch frames, and when the lettuce had been harvested the frames were removed, leaving the gladioli to flower at a later date. Runner beans were grown by the acre and were sent to market, much to the annoyance of the late dear old Ernie Cooper (more about him later), who thought the Institute should not be allowed to sell its produce. He saw it as unfair competition and wrote many letters on this subject to various horticultural magazines.

There was also a large area for strawberry, raspberry and blackcurrant production. I'm not sure who got the best deal out of these crops: the birds, the students or the ones who packed the fruit for market. The very worst job that one could be allotted was the winter pruning of the blackcurrants and raspberries, when the thick frost was still clinging to the bushes. It was on one such afternoon that I remember watching a particularly noisy aircraft flying overhead. It was the much acclaimed new Comet being tested out before going on to the production line at De Havilland, at nearby Hatfield.

The glasshouse department was my favourite practical location, as it fired my imagination and I soon became especially interested in the growing of tomatoes, carnations, roses and many pot plants. Somehow they just seemed more challenging than the more mundane market garden crops. We students enjoyed helping the glasshouse staff sterilise the soil before the tomatoes were planted in late January or early February. The work involved double digging and positioning steel pipes into the trench, before replacing the top soil. This had the effect of creating your next trench to be dug ready for the next pipe and so on, until eight pipes were buried with the top soil. At this point, the pipes were connected via a header to the steam boiler. A special rubber sheet was then placed over the top of the piped area and the steam was turned on. The temperature of the soil was brought up to 75°C and it 'cooked' for about two hours. During this time we were allowed a respite from our labours and had great fun telling stories and indulging in general student chitchat. Needless to say, some of these stories simply can't be put into print, but they certainly kept us smiling on those dull January days.

The carnation crop was grown in pre-formed concrete raised beds, which had the advantage of better drainage and were easier to sterilise between crops. The life of the carnation plants was normally three years, at the end of which disease would be starting to creep into the crop. So to avoid passing on diseases to the new crop, sterilization was essential. After the soil was prepared, a 6in square heavy gauge wire was placed on the beds and into each square a single plant was placed. About three weeks later, these plants would be stopped to create a bushy specimen and as the young growth developed, the wire frames were raised. More wire frames were added to support the plants over their three-year lifespan, by which time they would have reached a height of 6ft. Roses grew adjacent to the carnations and as they did not need support they were grown straight into the soil. Again, the average life of a rose tree, grown for a cut flower crop, was about three years.

Packing shed duty was always my favourite. The rose blooms were graded by the length of stem and were then bunched in dozens — in those days we had not been converted to the metric system. The carnations were marketed as single blooms and both crops were packed in large wooden boxes, their delicate heads being cushioned on paper spills.

The grading of tomatoes was done mechanically and unlike today, these were sold in 12lb boxes. At 5 o'clock in the afternoon, three times a week, the Institute's lorry made the journey to London, to the wholesaler George Monro Ltd. who was part of a very large concern in the horticultural world.

In the summer term, a coach took us to London's Covent Garden market at 4 o'clock in the morning. Here we had the opportunity to talk to the salesmen and to enquire about the quality of the produce we had sent the previous afternoon. We would ask whether it had travelled well and find out what prices had been reached. We would then retire to one of the many cafes, for a welcome breakfast, over which we would discuss the fairness of the sales.

The boys outnumbered the girls by about six to one, which made the social side a little unbalanced. Being a local lad, I coerced my St. Albans friends to send their sisters along to our regular social evenings. These events ranged from country dancing to quizzes. My best friend's sister eventually married one of my fellow students, so my efforts must have been appreciated. We also made forays to the various local colleges and I dare say we terrorised the poor unsuspecting girls. At one end of term dance, one or two of the likely lads decided to have fun by climbing the clock tower and crowning it with a student bicycle. I'll leave you to imagine the Principal's reaction the next morning — he was certainly not a particularly 'happy bunny'

... climbing the clock tower and crowning it with a student bicycle

Towards the end of my first year, it was announced that the Institute would be introducing a brand-new course, based solely on glasshouse production. I couldn't wait to enrol, as this was the subject that most interested me. The idea was to grow different plants from cyclamen, cinerarias and Primula obconica and to switch to the more fashionable indoor foliage plants, that were just starting to gain popularity. This was largely due to one Thomas Rochford, whose amazing nursery we also visited. It was at Turnford Hall near Cheshunt in Hertfordshire.

Thomas Rochford had a vision of English housewives adding a pot plant onto their weekly shopping list, just as their Dutch counterparts did. We grew a number of stock plants from which cuttings were taken. The young plants were potted into 10cm pots, grown on and then packed in mixed boxes of twelve to be sold at Covent Garden.

It was a great experience, as only six students applied for the course. We were taken to a wide variety of commercial glasshouse units and one visit that sticks in my mind — especially as it was only half a mile away from the Institute — was to a nursery that specialized in tomatoes. The owner showed us around and he drew our attention to the enormous stock of coal that was stored on the nursery to heat the boilers during the winter months. He reckoned there was enough coal there to keep the glasshouses heated for at least two years and I couldn't help reflecting on the enormous amount of cash that this tied up. My economics lectures must have had an influence on my way of thinking, or were my accountant Father's genes perhaps coming out in me?

Those 15 acres of wooden greenhouses have now been replaced with state-of-the-art, super-duper aluminium glasshouses, that produce tomatoes and cucumbers on a huge scale. Needless to say, they're no longer heated by coal, and they're no longer under the same ownership, either!

Towards the end of 1954 my parents, realizing that their only son had found his chosen path in life, suggested we could look for a nursery that would enable me to start a small business. To raise the necessary capital for this ambitious venture, my parents decided to sell their house in Marshalls Drive, since it had a valuable piece of land attached. As my Father was still employed by L. Rose & Co. Ltd., we decided to look around the county for nurseries with attached houses, which were within striking distance of St. Albans.

We viewed nurseries whose greenhouses were in any condition: good, bad and indifferent. In nearby Codicote, I remember inspecting a small nursery whose principal feature was a 20ft banana tree and a distinct lack of glasshouses. This didn't appeal to me at all, so after a fruitless search, we hit on the idea of developing a nursery from scratch and looking for a house nearby.

We heard of a forthcoming auction of some pasture land on the North Orbital Road, which was to be offered as a whole, or in five plots. With the excited anticipation of purchasing one of these plots, my Father I attended. We were bitterly disappointed when the entire 37 acres of land went as a single lot to an unknown buyer. Our search would have to begin all over again.

Soon after that, my Mother was showing a prospective buyer around our house and in the course of conversation, he asked why we were moving from Marshalls Drive. Mother explained the reasons and went on to say how disappointed her husband had been at the recent auction. Much to her amazement, he replied: 'I purchased those 37 acres of land for my firm, the Electrical Apparatus Company.' He went on to explain that E.A.C. wanted the back land for staff playing fields, but had little use for the land that fronted the North Orbital Road itself. He also told my Mother that if we wrote to him, E.A.C. might be interested in re-selling some seven acres.

Although he didn't buy our house, he left my Mother in a very excited state, clutching a piece of paper bearing the address that was to change our lives. I'll never know whether we'd simply enjoyed a lucky coincidence, or whether some guardian angel was watching over us. But I was over the moon and immediately started hatching the plans that would make my fortune. After all, hadn't I been taught that from x numbers of acres of glass, you could expect a return of y number of £?

I calculated how much heated glass I would need to start the magical ball rolling and with the profits from the first years I would soon have the whole of the 7½ acres covered with state-of-the-art greenhouses.

The entrance to the 7 acre parcel of land on the North Orbital Road
The insert shows the very same entrance today – just slightly larger

My parents decided to form a limited company, with my Father as Managing Director and my Mother as the Company Secretary and also a Director. Aylett Nurseries Ltd. was born. The first meeting of the directors took place at Marshalls Drive, with myself in attendance. It is noted in the company's minute book that 'Mrs. Aylett produced the Certificate of Incorporation dated the 29th July 1954. She also produced a print of the Memorandum and Articles of Association, as filed at Companies House.'

It was resolved that one fully paid ordinary share, numbered No.1, was allotted to Mr. Sydney Aylett, while No. 2 was allotted to Mrs. Muriel Aylett. It was also noted that, at the first annual general meeting, it was anticipated that Roger Aylett would be appointed as a director on reaching his majority.

My parents had two very dear friends, the Goldsmiths, whom they'd first met on their honeymoon. Because they shared similar interests and professions, they had continued to keep in very close contact with each other. We were delighted when Oliver Goldsmith agreed to become our accountant and auditor. He advised us for many years and we were able to take advantage of his vast knowledge of all things financial.

Aylett Nurseries Limited Share Certificate

Bankers were also appointed at this first meeting and an account was opened, in the name of Aylett Nurseries Ltd. at Lloyds Bank in Chequer Street, St. Albans. We still have the same bankers today, but obviously not our original bank manager Stanley France, who was such a huge help to us during those early years.

It was further noted that no remuneration would be paid to the officers of the company, but held in abeyance for the time being. I wonder if those two directors had any idea of what they were letting themselves in for – they certainly never received a penny in their lifetimes.

The land was acquired on 4th April 1955 and my studies at Oaklands were completed in July of the same year. Before signing the contract for the land, we contacted the St. Albans District Council planning department, setting out our 10 year projected proposal, which included the building of a number of glasshouses for the production of tomatoes, cucumbers and carnations. Our plans were duly accepted and we went ahead with the purchase. During the summer of 1955, we obtained planning permission for a packing shed measuring some 25ft by 12ft and a boiler house to provide heating for our three greenhouses, the largest of which measured some 140ft by 30ft and had been designed to grow carnations.

To my considerable disappointment, I have to tell you that in fifty years, we have never grown a single carnation on our site. Initially, we used that particular house for growing tomatoes and in due course it became a propagation house. These days, our furniture and Christmas showroom stand on the site.

During the Easter holiday, my parents had organised a very special 21st birthday party for me. It was held at the Water End Barn, an old structure that had been transported from the countryside near Wheathamstead and rebuilt in the centre of St. Albans prior to the war. This venue was popular with various organisations in the city, as it was able to cope with large numbers of people. It made a lovely setting for such an event. Little did I realize that I would in time be organising one of my own daughter's 21st birthday party there, followed by the Nursery's 25th anniversary and in due course its 40th. Sadly, our 50th anniversary celebration will not be able to continue the tradition, as this wonderful old building has now been closed and its future is uncertain at the time of going to print.

21st birthday greenhouse cake

My own 21st birthday party was a wonderful evening that remains in my memory as a very special event. My parents had a spectacular cake created in the form of a greenhouse with flowerpots as a border. For me, the evening also marked the culmination of two very happy years at college. I was joined by relations, school chums and fellow students, plus many other close friends from my youth. They all came to wish me well in my new venture and some even offered to help.

After that particularly unforgettable occasion, I was filled with great determination and enthusiasm to succeed and thereby to repay my parents for their unfailing belief in me.

There were a great many nurseries, both large and small, in the Lea Valley at that time. A huge diversity of commercial crops were grown under glass, the area was favoured partly because the quality of its natural light, which was believed to produce stronger plants and partly because of its easy access to the Covent Garden market, which was then all-powerful. As the need for residential land became more pressing, a great many of these nurseries closed down, selling off their old wooden greenhouses and moving down to the south coast, where the light was even better. They took with them the much-needed capital to invest in the newer aluminium glasshouses, which the Dutch were pioneering so effectively. I went to one of these sales in the Lea Valley and was successful in buying a 50ft by 10ft wooden greenhouse for the princely sum of £50.

I thought it would do very well for growing a crop of cucumbers and after dismantling it myself, I transported it back to our newly purchased land in St. Albans, where it was stored away until we had the necessary planning permission.

In the meantime, my parents had found a house in Mile House Lane, it suited us all ideally. My Father could easily get to his office, while I was well placed for the nursery. The only drawback to the house was its name, Sopwell Lodge. We were frequently confused with Sopwell House, which is today a luxurious hotel and country club, but was then an old peoples' home and our Sunday afternoons were often disturbed by would-be visitors in search of their elderly relatives.

During that Easter holiday I worked on a few schemes on the Nursery that I felt would fit in with my course work during my final term at Oaklands. I certainly didn't want to jeopardize my final certificate and sacrifice my hard work, by not studying as hard as I possibly could. One of these projects involved felling a tree and cutting up the timber to clear a piece of land which would be suitable for growing the few dahlia tubers that I had stored from the garden at Marshalls Drive. After clearing this plot, on what I call the backfield, I was able to arrange for Dennis Holmes, our local agricultural contractor, to plough it for me. I then planted a few chrysanthemums, to be grown alongside the dahlias as a cut flower crop.

The holidays came to an end and it was time for me to put my head down again and work hard at my studies. I didn't have much time for my new projects during that term, but I did manage to squeeze in a few hours at the weekends and my loyal student friends lent a hand with the basic jobs, such as keeping the plot

free of weeds. They also helped with the disbudding of the dahlias and chrysanthemums, to ensure that I had a good chance of being able to produce some flowers that would be of marketable quality.

Of course, I didn't employ any staff at this time. My Father, realising that we needed to do a certain amount of land clearance before his son's ambitious plans could be implemented, suggested that students, being students, might be in need of some funds and consequently willing to undertake some of the vital menial jobs, such as digging in rabbit wire to keep the little perishers away. Rabbits seemed to be particularly partial to dahlias for their lunch.

DAHLIAS THIS WAY
Thank you

... Rabbits seemed to be particularly partial to dahlias for their lunch

At the end of term, one of these students asked if he could join me, as he'd enjoyed the work and was fascinated by the project. His name was Fred Le Mage and he became our first full time member of staff. Not being a local man and needing accommodation for his wife and himself, he decided to buy a caravan. We obtained the necessary planning permission and it was sited at the back of the nursery. Fred remained with me for a number of years until he decided, after the birth of his first child, that he needed a house of bricks and mortar. So he found himself a new job that provided one.

As for me, I had completed my two years at Oaklands and left the Institute full of enthusiasm and armed with my certificates, that declared me to be 'proficient in general and glasshouse horticulture.'

... the answer was to create a retail selling area

The dahlias that I had brought with me from Marshalls Drive looked lost in my newly purchased land, so I managed to buy another 200 plants from Ernie Cooper, a local commercial dahlia grower, who cultivated many acres of them for cut flowers and was an exceptionally gifted hybridizer. He was a well-educated man and had studied at Kew, but being somewhat outspoken, he was never quite at home with the establishment. He was really most eccentric and dabbled with many species, including canaries and runner beans and it was he who bred and introduced the famous 'Jescot' strain of dahlias.

Mr. Cooper was interested in why I should be buying a quantity of dahlias and I explained that I was setting up my own nursery on the other side of town, near London Colney. He was very kind, showing a great interest in my venture and was able to suggest some varieties that would have the qualities that would make a good cut flower.

When I went back to collect my plants, Mr. Cooper asked me if I had any spare land on which he could raise some dahlias of his own. He made me an interesting offer, if I could get the land ploughed and ready by June he would plant three acres of dahlias. He would pay me no rent, but I could harvest the flowers and keep the revenue for myself. It seemed an admirable arrangement. I had spare land that was idle and all I had to do was get it ploughed and cultivated, so on a handshake this was agreed. I spoke to the principal at Oaklands and he agreed that I could borrow the tractor as long as I paid the driver, Mr. Harvey, for his work. He ploughed the three acres and Mr. Cooper rotovated the ploughed land

and planted the young dahlias in early June. He and his team then continued with their cultivation. The dahlias flourished in the virgin soil, and we were expecting a bumper crop of flowers. I made up a sample bunch and hawked them around Watford and St. Albans, calling at florists as well as any greengrocers that I thought might be able to sell them. I didn't always get the warmest reception, but I was not deterred and I persevered until I had the promise of some prospective sales.

Realising that I couldn't bunch the flowers by myself, I advertised for some part time staff. I was lucky and took on to the payroll two part time ladies named Olive Kemp and Ada Crain. They were paid two shillings and three pence an hour, which is equivalent to approximately 11p in decimal currency. Olive and Ada stayed with me for many years as very loyal employees, working during the spring and summer and glad to have a rest during the winter months.

The following year I recruited three more part time ladies and so my little workforce grew. I also needed suitable transport for delivering the flowers, so I bought a second-hand Wolsey car with a large boot. To give me an increased payload, I removed the rear seat and attached a roof rack.

We spent the summer bunching dahlias and in September we moved on to the chrysanthemums. At the end of the season Mr. Cooper came with his staff and lifted the tubers and took them back to his nursery. He was well satisfied with his half of the bargain and I with mine.

During that summer of 1955, we had to build a roadway so that we could access the land. I don't recall where I hired the necessary equipment, but I do remember a special earth-moving machine that was used to remove the topsoil. This was retained in a large heap for future use, as our loam was an important ingredient in the John Innes compost I intended to mix. It was a valuable asset for the next ten years and I was very pleased that we had not had it carted away. Once the topsoil had been removed, we backfilled the proposed road with hardcore and topped it off with hogging.

I had noticed some contractors working outside, on the North Orbital Road. They were using just what I needed to finish my road, namely a mechanical roller. So during the lunch hour one day, I approached the boss man and negotiated with him to roll my new road. The job was done after work one evening. I'm perfectly certain this was quite irregular and against all rules. But life was much simpler in those days and a free pint of beer could move mountains.

Earth moving machine

We also installed the essential services that summer, starting with the electricity. The gentleman from the electricity board who came to see me was very helpful and advised me on the size of the supply cable. He told me that the cable he recommended was the same size as that used by Marks & Spencer and he couldn't foresee that we would ever outgrow it. In fact, we have had to replace the cable twice since then. We purchased a small 8ft by 6ft shed to house our electricity meter and into this the electricity board installed the large cable.

The next requirement was to be able to communicate with the outside world and to have a telephone installed, so I bought another, slightly larger shed and we had it put up at the front of the nursery. The GPO erected the telegraph pole and installed the phone. The shed became my first office and now I could feel that I was really in business. My Mother was my secretary, but as she wasn't too keen on working in a shed, she preferred to do my letters at home.

While we were living in Marshalls Drive, we had become very friendly with our next-door neighbours. Mr. Fisher was the managing director of a local printing business, so I went to him to seek his advice on my corporate image. He designed the headed notepaper that developed into the logo used in all our printed stationery. This same logo was used as our first sign, on the front of the nursery and had to undergo some stringent planning scrutiny. Consideration was even given to the shrubs that were planted beneath it and I seem to remember that we bought some forsythia for this purpose, as the yellow flowers blended in well with the yellow of the sign.

The Water Board installed a 3in water meter and a local builder laid the 3in main, with stand pipes at equal points through the centre of the land. The builder also constructed a much-needed brick building to house the two toilets, as we had no mains drainage and consequently no sewer, he also had to construct a septic tank.

I needed somewhere secure where I could keep my tools and also a building that was capable of being used as a packing shed for the cut flowers I intended to grow. So I found a local source that sold prefabricated garages and selected one that would fit the bill. It had a large metal framework, with asbestos panels and windows along one side, in front of which we constructed a bench for my bunching girls to work at. We painted this building a darker shade of yellow to continue with the corporate image.

Dahlia Fields from the North Orbital Road   Packing Shed

The small greenhouse, measuring 25ft by 10ft, that we had in Marshalls Drive, had been taken down and rebuilt on the right-hand side of the roadway towards the top of the nursery. This was our only heated greenhouse during that first winter; it had a small coal-fired water boiler that needed frequent stoking. We used it for propagating our first dahlia cuttings, which we potted up to create our stock plants for planting out.

The dahlias created quite a spectacle from the North Orbital Road and people passing by often stopped and asked if they could buy a bunch. It seemed ironic that I was having to work so hard to sell my bunches of dahlias, while here was a self-made opportunity, where people were begging me to let them buy some flowers. The answer was to create a retail selling area. Funds where very short indeed, so I ventured down to the local scrap merchant where, for the princely sum of five shillings, I purchased an iron trestle along with a quantity of glazed earthenware jars, which were also for sale at a bargain price. Furnished with these items, I created at the entrance to the nursery a display of dahlias that attracted many regular customers, who would stop on their way home from business to buy a bunch for their wives.

The first retail outlet

I then read in the local paper about the forthcoming sale of an old nursery near Knebworth. I decided to go along to view the various lots, to see if there was anything that I could utilize to increase my productivity of cut flowers.

There was a quantity of cold frames — commonly known as Dutch lights—that caught my eye. I was lucky at the auction, being able to purchase them for a very modest sum. I got a local carrier to transport them back to the nursery and then bought some 9in by 1in timber boards, treated them with Cuprinol and . . . bingo! I was the proud owner of a number of cold frames.

The frames I then planted with gladioli corms and I germinated some lettuce seedlings that I planted between the rows. The lettuce matured first and I marketed these, leaving the gladioli to develop. As they budded and grew, the cold frames were removed, leaving the gladioli to flower. These I managed to sell to local florists as cut flowers.

Lettuce being harvested from the Dutch light frames.
In the background the Hortus Dutch light structure (see p52)

During the short winter days of 1955/56, we had double dug a piece of land between my office and the packing shed. We erected wooden posts to support a framework of 6ft bamboo canes, with the idea of growing sweet peas during the coming summer. We sowed the seeds in the spring and having grown sweet peas as a teenager, I just followed the same technique, but on a larger scale. The crop proved an outstanding success and I had little trouble finding sales outlets, as they were very popular with local florists. These entirely different flower crops helped with the financing, but were far from being enough to provide me with a living wage. My Father was still supporting me financially and of course I was living rent-free at Sopwell Lodge.

In my early years in business, one of the very first things I realized was that I hadn't given sufficient thought to the fact that in acquiring a piece of maiden land and trying to develop a nursery from scratch, I had to go out and buy all the very basic essentials that I needed, everything from a screw to a piece of string. I must admit there were times when I thought it would have been more prudent to have purchased an existing business, as these necessities soon ate into my diminishing capital. I very soon learnt to make do and mend and developed a reputation for creating my own special brand of improvised DIY, using telephone wire and 6in nails.

Later in the autumn of that year, we started the preparation work that was necessary before we could install the two boilers. This included the digging of an area of 30ft by 20ft, to a depth of 5ft. Unfortunately, before we could finish this task, the weather broke and the hole filled with water and later with snow. The plans that had been passed by the local planning department included an acre of heated glass. At the same time as the boiler house footings were being dug, the first of the Hortus carnation greenhouses, measuring 140ft by 30ft, was under construction. I say 'the first', because it was all I could afford at that time. I planned to grow carnations, but I needed funds to construct the raised beds of concrete and my capital was running low, so in the interim I decided to grow a crop of tomatoes.

It was my intention to construct one greenhouse per year until I had filled one acre. The funding of this project would come from the profits from the previous year's crop. I certainly had my rose-coloured glasses on when I envisaged that scheme, as it would actually be some fifteen years before I was able to declare a profit. I did manage to scrape enough money together to have the footings and brickwork base completed, so that I could erect the bargain greenhouse that I had bought the previous Easter and had been storing.

Boilerhouse in snow

It was decided that it was prudent to purchase a similar greenhouse from Hortus while they were still on site, as it would cost us considerably more to call them back again later and we could see that we would need more propagation area the coming spring.

It wasn't until spring of 1956 that the boilers were installed. The boiler house was then built over the top of them. The boilers were second-hand and the man who installed them had managed to find them at a good price. In those days a gravity system was used to heat the greenhouses, in other words the boilers were at the lowest level and the pipes rose 1in every 10ft. This meant that the pipes at the furthest point of the carnation house were some 3ft off the ground. The two tubular boilers were coal-fired, a lorry delivered the coal every four to six weeks depending on the weather. Ten tons of coal, which was one delivery, cost £143.00. We shovelled the coal down a ramp into the boiler house, it was certainly backbreaking work.

Boiler base and building

In April 1956 a new addition to the team arrived. This was a young man aged 15, who had just left school. His name was Brian Fowler and he became my second full time employee. Next year, 2006, he will have completed fifty years' service with the company.

After Brian had definitely decided that horticulture was for him, he too attended a full time course at Oaklands and gained his certificate. On his return, his special responsibility was for the dahlias and a number of years later he became nursery manager. Today, Brian is site manager and still continues to look after the production of our dahlias.

The Carnation house – note the chimneys under construction for the boilers

Tomato crop growing in the finished greenhouse

1956 was a memorable year. It saw our first tomato crop, which was very successful. The crop was unusually heavy, being equivalent to 50 tonnes of tomatoes to the acre, which in those days was considered above average. I could almost see my second greenhouse!

Our 7 acres were very exposed and there was an airfield and factory close by belonging to Hawker Siddley, where they used to test fly their new aircraft. The wind whistled down the road and to help counteract this phenomenon, I came up with the idea of growing runner beans between the rows of dahlias. So I hired a tractor with an augur behind, which drilled holes some 2ft deep, into these we dropped 8ft rustic polls, which were joined together using our multi-tasking telephone wire. The runner beans grew well up the strings that we tied to this framework of wire.

You may be wondering how I came to be in possession of so much telephone wire? I bought it after answering a government surplus store advertisement in The Grower magazine. We had a splendid crop of runner beans and I remember thinking that, as they were very expensive in the shops, we were going to make a lot of money. When the first consignment was ready, I sent them to market with great excitement. I telephoned the next morning to see how the runner beans had sold and you can imagine my disappointment at being told that none of them had actually sold that morning, although they had great hopes for tomorrow.

The next morning I phoned again and they still hadn't sold and a week later the Westminster Council sent me a bill for disposing of my wonderful, but now rotten, runner beans. This was my first lesson in not assuming that I could sell anything I grew and that there was no long queue just waiting to purchase Aylett Nurseries' products.

After the previous year's experience, I had decided to increase my own dahlia production, but as they came into flower I experienced difficulty in increasing my selling outlets, so I started sending them to Covent Garden market. The Wolsey had disintegrated by this time and had been replaced with a Standard Station Wagon, but this was no use in the transportation of heavy wooden flower boxes. So I enlisted the assistance of a man with a lorry, who had a business transporting growers' produce on a nightly basis to the market.

He collected our boxes at 5 o'clock in the afternoon, on five days of the week. The following morning we would phone the wholesaler to find out the price the previous night's sending had made, often to be told they were still on the stand unsold. The market was definitely not living up to my expectations. There was nothing wrong with our flowers, but I soon realized that it takes years to establish a rapport with the wholesaler and I was a newcomer.

One day in 1957, I was complaining to a representative from Murphy's Chemicals that nobody seemed to want to buy my dahlia flowers. He said 'Why don't you bring some of those dahlias to my local flower show at Harpenden? The show attracts plenty of visitors, who show a particular interest in our trade stands.'

So I bought some containers and Mother and I went to Harpenden Public Hall to arrange our exhibit. When the show opened, the first thing we were asked for was a catalogue and I'm afraid I told a little white lie, saying that it had been held up at the printers and that if they would leave their name and address we would send them one in the post.

On my return from the show, I quickly contacted a printing firm and they rushed through a basic list for me. This was the first of many shows that we staged in Harpenden and it was our introduction to the world of the mail order business.

My Mother Muriel Aylett at Harpenden Public Hall

The glass house manufacturer Hortus, who built the carnation house that we were growing tomatoes in, came to me saying they had designed a structure to make better use of Dutch lights, in effect turning Dutch lights into a greenhouse which obviously had better growing potential. They thought that our site, being situated adjacent to a main road, was an ideal location on which to erect a demonstration house. As I mentioned earlier, we had purchased a number of these lights at a sale and although we had made use of several of them, we had by no means exhausted our supply. Hortus offered to provide and erect the structure free of charge. All we had to do was fit in the lights and allow them to have access to the site so that prospective buyers could come and view their new design.

It was an offer I could not refuse and the deal was clinched. Making use of our Dutch lights in this way made much more sense, as it gave us a cold greenhouse in which to grow tomatoes during the summer months and then to follow on with a cut flower crop of chrysanthemums. Although we were very pleased with our part of the bargain, to my knowledge Hortus' new design was not taken up by many growers and in fact not a single grower came to look at it.

Roger Aylett outside the tomato house 1957

Setting up a business takes capital, that's fundamental, but I made the mistake of thinking that my working capital would be financed by my profits. A horticultural business that starts from scratch needs more than my beloved Father had envisaged and we were running out of funds. I had two dear aunts who, on hearing of my dilemma, offered to buy some shares. Although I'm sure they had little hope of seeing any dividends, their help and their belief in me kept the nursery solvent.

One morning the local careers officer telephoned me to enquire if I would be interested in having a student to work for me. They had a young lady who lived in Hendon who wanted to make a career in horticulture. I had no objection to this proposal, as having had all the encouragement in the world to take up my own chosen career, I liked to be in the position of helping others do the same. The young lady's name was Hazel Rowland and she arrived with her Mother for an interview. They had travelled by train to St. Albans and then by the No. 84 bus. I conducted the interview in my office and then showed them around the nursery. Hazel's family were keen gardeners. Her Father grew prize chrysanthemums and her Mother was a very dedicated flower arranger. After the interview was over and they had left, I remember asking myself why she should want to come and work on a small nursery like ours.

Hazel joined us in mid-August of 1957. She arrived every morning at 9 o'clock and worked until 5 o'clock. In those days, students didn't expect to be paid a salary, but just to receive some pocket money. She tells me that this was the grand sum of £1 pound per week and that her train fare was eleven shillings per week.

She was very enthusiastic and tackled any jobs that cropped up, such as cleaning the tomato house out, taking dahlia cuttings, stoking boilers or mixing soil. The latter incidentally, was all done by hand and pushed away in wheelbarrows. It was quite a few years later that we mechanised this job and bought a cement mixer to take over a very backbreaking task. We left Hazel in charge of the nursery one day, while the male members of staff went to prune some fruit trees in a customer's garden. She seemed a little bit stressed when we arrived back. We found that the boiler expansion tank had been overflowing and that she

had had to climb on top of the boilers to tie the ball cock up.

It was in the September of that year that Hazel accompanied my Mother and I, when we staged our first dahlia exhibit at the Royal Horticultural Society. Years later she told me that she didn't enjoy her day, as Mother and I couldn't make up our minds where the arranged dahlias looked best and she spent her whole day moving vases from one place to another. Very soon her year came to an end and Hazel left the nursery to start her studies at Oaklands. We had become good friends during her year with me and our friendship blossomed while she completed her studies.

On the day she left Oaklands, having successfully finished her horticultural course, we became engaged to be married. She returned to the nursery with the intention of organising the proposed flower shop.

One of our first exhibits of dahlias – note the baskets that they are arranged in are those my mother taught Hazel to make

chapter 5

... august 1961 awarded our first gold medal

In 1958 we made two momentous decisions, both really brought about by dahlias. At the same time as the markets were returning terrible prices for my flowers, my retail sales were growing, both in cut flowers and plants. Dahlias were the star attraction and we were selling more and more of them to the passing trade. We therefore decided to specialize in dahlia production in a big way. This was before the days of garden centres, where you can now purchase a vast selection of different species of plants. In those days, nurserymen specialized in a particular genus. They exhibited at flower shows around the country to fill their order books, the orders taken were despatched to their customers the following year.

The second momentous decision we made, was to build a shop on the front of the nursery to enable us to sell more flowers. I heard of a small chicken shed measuring approximately 25ft by 15ft, which I thought could be adapted into a building suitable for a flower shop. It was going for a song, which suited my somewhat limited budget. Planning permission was obtained very quickly to erect this shed, on piers some 2ft high, so that it was on the same level as the road.

A friend put me in touch with a handyman, who reconstructed the shed into my first shop. He lined the building inside with hardboard, making a partition to one side that created a tiny room that we intended to be used as our second office. He then glazed the front with some of my magic Dutch lights, installed a telephone extension and the water and electricity were both connected.

My Mother contacted Margaret Best, a fully qualified Constance Spry florist, with the idea of having some private lessons in floristry. This would enable us to execute bridal work and funeral tributes. I applied to Interflora to join their worldwide organisation, but was turned down that first time. We reapplied some time later and our membership was accepted in 1960.

The first garden shop at Aylett Nurseries

chapter 5

Hazel had a friend, who had been on the course at Oaklands with her, who was looking for a job that would let her stay in St. Albans. Hazel felt that she was capable of serving the general public and although she would really like a job on the nursery, she was quite happy to help out in the shop when required. So Penny joined the team and found lodgings with one of our part-time ladies in London Colney.

Hazel had been brought up with flowers and having inherited her Mother's natural talent in the art of flower arranging, it was particularly easy for my Mother to teach Hazel floristry. They spent many happy hours together creating various designs. The other vital skill that my Mother imparted to Hazel, was the making of wicker baskets to exhibit our dahlias in, since those early exhibits showed as much of the container as the flowers themselves, the baskets needed to be as aesthetically pleasing as possible.

Although Hazel and I had done a basic bookkeeping course at Oaklands, we needed further coaching in this subject to be able to provide sets of figures for the accountants. Father made an admirable tutor, teaching both of us the finer points of a bought ledger and instilling into our heads the necessity of keeping accurate figures.

The shop began to build up a loyal following and we gradually introduced various garden sundries that fitted into our small space. The opening hours in those first years were 9 o'clock to 5.30, six days a week. We were initially closed all day Sunday, but eventually we opened on Sunday mornings too. These were the shop hours; the nursery hours were a different kettle of fish and we were often to be found working late into the evening.

I mentioned earlier that we had exhibited for the first time in 1957 at one of the Royal Horticultural Society's fortnightly shows at Vincent Square in London. At this show we had learnt a few lessons, one being that we were very much the new boys and we had to work our apprenticeship. There was certainly no shortage of dahlia growers to compete with. In 1958 it was decided that we would exhibit at the National Dahlia Society show in London, three of the R.H.S. shows in August and September and at Reading, Leicester and Shrewsbury.

My memory is rather hazy as to the exact dates that two important members of staff joined the team, but I do know that it was about this time that Fred left, and Ralph Skuce arrived to replace him. Then William Saunders came to see me. He was a friend of mine from schooldays at St. Georges and had just completed his course at Oaklands. He said he would like to come and work with me to help develop the nursery. This proved to be a long-term commitment and he stayed for many years as one of my stalwarts.

It was while William was with us that he struck up a friendship with a young lady that worked alongside Hazel. Doreen could turn her hand to almost any task, and was another valuable member of the team that constructed our stands at different flower shows. Their friendship flourished into marriage and they eventually left us to set up their own business. This didn't involve horticulture: William and Doreen ran the village store and post office in a village in Kent. The village's gain was our loss, for we missed their humour and company.

For transporting our flowers to those early shows, we bought a two-wheeled trailer that was towed by our newly purchased Land Rover. I had managed to write off the Standard Station Waggon after a late night out. The dahlias were packed into the wooden market boxes, loaded into the trailer and securely tied. I had picked up quite a few tips about using the right kind of knot from our market carrier. When we arrived at the showground the first job was to unpack the dahlias and plunge them into deep water, ready to be arranged.

In those early days we used scrunched up chicken wire to hold the flowers in position. We didn't have a pre-designed plan to work to and arranged our flowers on makeshift benches that we constructed out of the empty boxes. The search for water was always fun and we carted the heavy buckets for what seemed like miles.

While Hazel and I arranged the dahlias, William would put the stand together under my instruction. This could be a very Heath Robinson affair depending on the show. Some of the stands were extremely precarious and certainly wouldn't pass any health and safety checks today.

...under my instruction. This could be a very Heath Robinson affair

As the dahlia arrangements were completed, they were handed up to William for placing on their allocated pedestals. As we progressed we might then decide that we didn't like that particular dahlia in that position and would move it. You can imagine how long this torturous process took and we hardly ever finished an exhibit until the early hours of the next morning.

Our favourite show, without a doubt, was Shrewsbury. There were at least five marquees full of horticultural produce, one of which housed the most exquisite exhibits, from what must have been very grand private gardens. These exhibits would contain plates of mouth-watering peaches, nectarines, strawberries, melons and grapes, all displayed very tastefully This show was – and still is – held in the Quarry Gardens alongside the River Severn. The gardens are particularly lovely, and the people most welcoming. The show is organised by the Shropshire Horticultural Society and the park superintendent in those days was Percy Thrower. What a gentleman he was! He always had a word with every single exhibitor, and as newcomers this certainly meant a lot to us. I can see him now, walking the showground in the early hours of the morning, with his black Labrador at his heels, offering encouragement and making sure we had everything we needed.

At Shrewsbury you had to win your spurs before you were allowed to exhibit in the Quarry Marquee. This was reserved for the crème de la crème and needless to say we were not exhibiting there. We couldn't find rooms in the town so we borrowed a tent and slept with the friends that we had managed to coerce into coming with us. This partly was so they could chaperone Hazel and myself, and partly so they could lend a much-needed hand.

My memory fails me as to exactly what awards we won during that initial year, but I know we won a Gold Diploma at the Reading Show, as I still have the card in my files.

The awards differed from show to show. We were not charged a fee for exhibiting because, after all, our exhibits were the show. The judging took place in the early hours of the opening day and the top award would be a Large Gold Medal. If you were very lucky, you might also win the Best Trade Exhibit, which often carried a special prize. Then there came the Gold Medal, followed by Silver Gilt and Silver and then Bronze. But these were medals in name only: instead of the actual medal, a cash sum was given, and the higher the medal, the more cash you received. A gold medal was worth somewhere in the region of £100. The R.H.S. was the exception to this convention, in that it awarded actual medals.

Our first Chelsea Flower Show exhibit

Another bonus was that at the end of a show, a sort of mad fever gripped people and they clamoured to buy the dahlias. These would be almost dead by this stage, but who were we to refuse to sell them? We often made a tidy sum from these sales. The revenue helped to finance our showing expenses, but we were dependent on being awarded a good medal.

We were quite pleased with our achievements in the show world and in the following year, 1959, we decided that we would apply to the R.H.S. to show our dahlias at the prodigious Chelsea Flower Show. One of the pre-requisites was that you had exhibited at their Vincent Square shows, so at least we qualified on that count. Dahlias do not flower naturally in May and I think it was because there was only one other grower exhibiting the genus at Chelsea that our application was accepted. The dahlia tubers were started off in January. We decided to grow them in some wooden boxes that we had manage to acquire from a local greengrocer, as we thought that these would be easy to transport and would fit together to make up the full display, which was to take the form of a dahlia garden.

We grew these dahlias in the little greenhouse that had come from Marshalls Drive. We nurtured them like babies and they rewarded us by flowering on time. The great day came and the dahlias were transported to London and placed carefully together to create our display garden. We edged the garden with grass that we had grown on old felt, using Timothy grass seed that has the characteristics of germinating quickly to make a thick greensward.

My Mother and a team of loyal family and friends that I had managed to persuade to help, looked after the exhibit. We were so busy that we needed a representative on each corner to deal with customers and we came away from that show with bulging order books. I think what attracted people to our dahlias was the fact that we promised delivery the following week and that was a promise we kept. Another great friend of ours, Merle Bacon, who helped me out from time to time, collated all the orders and prepared the necessary paper work. We had made it successfully through our first Chelsea Flower Show and we resolved to return the following year.

We decided to improve our trailer, so that we could transport more equipment and hence enter the larger trade stands class. This would give us a chance of winning the coveted Large Gold Medal, with its enticing cash prize. William and I constructed a plywood box, using Dexion for strengthening and we fitted this over the trailer and bolted it in. We painted it yellow to fit in with our corporate image and into this contraption we loaded our show equipment.

At the same time we organised our carrier to bring our boxes of dahlias to the various show grounds. He would meet us at the showground, often after we had offloaded the equipment and set up the stand. Later on we changed the boxes for plastic buckets to transport our prize blooms and to save time putting them in water when we got to the show. We had given up arranging the dahlias in chicken wire, as it was much too slow and there was now a product on the market called Flora Pack, which you crumbled up and soaked and then packed into the containers. Unfortunately, it was very expensive so we reckoned that working on the same principle you could use peat as a substitute. This worked a treat, the only downside being that it was extremely heavy to transport. One consequence of this was a near disaster, that struck one year when we were on our way back from Shrewsbury, on the A5 just outside Coventry. All of a sudden, there was a loud bang and the trailer's tyre gave up the ghost. We scrambled out to view the damage and realizing that we were well overweight, decided to give the hedgerows a treat and dumped the peat hastily before heading for the phone box to summon assistance from a tyre specialist. How we got away with not being pulled over by the police I will never know. The peat was also extremely messy to prepare, but it was cheap and we continued with this method for many years.

The other change to our exhibition strategy was that we finally gave up on trying to mix up the colours. It just didn't work for us, although other exhibitors seemed quite happy to put scarlets next to purples. I have to say that the judges didn't seem to care whether the colours clashed, or just plain shouted at each other. We on the other hand, seemed to spend ages moving a bowl about because the colours didn't go with the bowl adjacent to it. This was perhaps because dahlias are such strong colours.

Hazel, being very colour conscious indeed, came up with the idea of colour grading the stand. We started at one end with scarlet varieties, blending through to orange, salmon, yellow and white, which then led into soft pink, deep pink, lilac, purple and we finished up at the opposite end with crimson. This method was popular with our customers, as they found it much easier to make their choice when all the varieties of one colour were grouped together. The judges took a little longer to come to terms with this new way of exhibiting, but we liked it and we never arranged a mixed stand again.

We got very tired of relying on the shows' catering facilities and decided to become self-sufficient. We were wasting a good deal of time queuing for mediocre food and concluded that it was time to do our own cooking. The answer was to make further use of our boxed-in trailer, so after we had unloaded it we would set about building ourselves a makeshift kitchen out of the dahlia boxes. We purchased a Calor gas hob, complete with grill and appointed a cook, usually Hazel, whose job was to create some yummy food. The other exhibitors thought we were a very funny lot from St. Albans, but I suspect were slightly envious when the smell of freshly cooked bacon and eggs reached their nostrils early in the morning.

... slightly envious when the smell of freshly cooked bacon and eggs

We needed more staff to help us construct the exhibits and it was about that time another young man came to see me who wanted to earn some money in his vacations from the theological college where he was studying. His fiancée lived just up the road from the nursery and she had suggested that we might require some casual labour. His name was Peter Hill, but we called him Pete the Parson and indeed, he went on to become a real vicar and later a canon. He is a great guy and his banter with William would keep us amused for hours.

The following year the weather was much colder and the dahlias we grew in boxes for Chelsea failed to flower on time. I spoke to Oaklands, who had some special orange lights that they used to propagate cucumber plants under and a little cheekily, I asked if they had any spare space available for our dahlias for the last week before the show. The dahlias were transported to the Oaklands glasshouse, where the light treatment didn't prove to be all that effective in forcing the number of blooms that we needed to make an impact. After the previous year's success and with the certainty that cancelling our allotted space would create a very black mark on our record, I came up with an idea and got in touch with a friend of mine. His Father had a nursery near Milton Keynes, and grew dahlias under glass as cut flowers for an early cash crop. I was relieved to find that although his crop was not in full flower, he had a small quantity that I could purchase. I collected these flowers from him the day before the show and tied them onto our dahlia plants by putting them in test tubes, making sure that these were well concealed by the foliage. The exhibit was not as colourful as our previous one, but at least we were there and we filled our order books again.

We had been exhibiting at the Royal Horticultural Society's shows since 1958, and in August 1961 we were awarded our first gold medal. Some celebrating certainly took place that evening! We were immensely proud of achieving this and as far as we were concerned, we had arrived. Later in that year we were awarded two more gold medals from the society. By the end of the 90's we had been awarded an R.H.S. gold medal each year for 36 consecutive years.

Our first R.H.S. Gold Medal

... more glass to grow the quantity of dahlias envisaged

As a result of our specialisation, we needed more glass to grow the quantity of dahlias that I envisaged selling. I had run out of money; all the funds were diminished and our turnover was not sufficient to generate any spare cash. I was also a long way off declaring a profit and I remember my Father likening the financial progression of the business to a never-ending process of throwing money into a bottomless pit. That is not to say that he had withdrawn his support; on the contrary, both my parents couldn't do enough to help me. They just desperately wanted the business to be viable.

In the spring of 1959 I heard of a sale taking place on the other side of Uxbridge. On offer were 20 acres of glasshouses constructed of English lights, which differed from Dutch lights by the size of the frames. Dutch lights consisted of one sheet of

English lights metal framework

glass, while the English ones had many smaller sheets. The whole nursery was to be demolished to make way for the construction of the Brunel University, all the buildings, fixtures and fittings were to go under the hammer. Ten days before the sale I went to view, with the idea of solving my glasshouse shortage problem. I was utterly amazed by the sheer size of the site and the intriguing design of the glasshouses. I had never seen glass constructed in this way and it certainly never featured in our college textbooks.

On my return I spoke to my Father, to see if he thought we could raise the necessary capital to enable us to purchase a half-acre block of the glasshouses. After considerable discussion, we decided to approach a close friend of the family who had just retired and with whom my Father and Mother regularly played cards. Mr. Peasley had always been interested in our venture and Father thought he might like to make an investment. Happily for us this was indeed the case, and we took out a mortgage with him for £8,000. He also bought a number of shares in the nursery, this injection of cash enabled me to purchase the half-acre of English light houses for a minimal sum.

That was the easy part. We then faced the daunting job of dismantling the glasshouses, which were constructed with steel arches that supported a wooden gutter. The arches were substantially concreted into the ground and proved impossible to dig out by hand. A very helpful gentleman, who had bought five times the acreage of glass that I had, who was a lettuce grower in that area, had organised the hiring of a crane and a pneumatic drill. He generously offered to let us muscle in on his hire contract in return for a small contribution.

William and Ralph accompanied me to Uxbridge and we spent two weeks of back aching work dismantling the glasshouses ready to be transported back to St. Albans. When we had finished, we had just over thirty tonnes of equipment, which was loaded into hired lorries and ferried back to the nursery over the next three days. A grower friend of mine recommended a glasshouse builder who could reconstruct the glass. This builder came from Hoddesdon in Hertfordshire, and had a vast experience of building commercial glasshouses. His name was Mr. Jim Salmon and he spent most of that summer with us.

Roger and Hazel in front of the completed glass house constructed from English lights, note the straw bales at the base

a distant dream

We now needed to install the heating pipes. Mr. Rafferty, who was one of the market carriers we used, had many contacts in the Lea Valley as a result of his market round. Hearing of my quest to find some cheap heating pipes, he came to me saying he had found a nurseryman who was pulling out his 4in cast iron heating pipes, to replace them with the more efficient $1\frac{1}{4}$ inch steel piping. A deal was done and Mr. Rafferty transported the pipes back to our nursery, where they were then installed. Funds were once more getting low and we still had a gap round the side of the English light house, so as a temporary measure, we made do with straw bales but unfortunately these became infested with red spider mite, which then spread onto the dahlias. So much for saving money . . .

The business was now based on the production of dahlias. The carnation crop was a distant dream and the tomatoes had come to a sorry end. We set about converting the tomato house into a dahlia propagation unit by building benches over the heating pipes and concreting the paths. It was on these benches that the dahlia tubers were bedded out in late December and the first cuttings were taken by the end of January. We had read in the horticultural press about a revolutionary misting system that had proved very successful in speeding up the rooting time. McPenny's, who were based at Bransgore, in Hampshire, had developed the system and we went to visit them. As a result, we bought enough equipment to complete an 80ft rooting bench. The system worked on the principle that as the solenoid valve dried out, the very fine mist cut in, thus reducing transpiration and stress and creating a suitably humid environment.

Unfortunately the valves had to be kept spotlessly clean in order to operate correctly. If there was any lime scale build-up they were liable to malfunction. I well remember coming in one morning to find a portion of the bench, complete with dahlias on the floor, as the solenoid had stuck and the weight of the water was simply too much for the bench. I can't remember how many years we kept this system, but it was never a great success as the cuttings developed botrytis and damped off, so we adopted the method of rooting directly into the bench, using perlite as our rooting medium.

Another development that took place in early 60's was that Mr. Cooper was approached by the city council to sell his land on the other side of St. Albans. They made him a very generous offer and as he was coming up to retirement age, he decided to accept. He found an ideal property just outside Bognor Regis, where he would have plenty of room to develop his breeding of 'Jescot' dahlias and his wife would be able to follow her love of piano music. It demonstrates his eccentricity when I tell you that he bought her not one but three grand pianos. He offered to sell Aylett Nurseries his extensive mailing list and we accepted. Mr. Cooper also sold us a small prefabricated building, consisting of half-glazed metal panels, which we thought we could use to construct a potting shed, to be sited at the end of the English light house. Having transported the panels across St. Albans, we bolted them together, found some timbers to make a roof, covered them with roofing felt and thereby acquired a perfectly adequate potting shed that cost us next to nothing.

We felt it was a little chilly for the ladies who sat potting the dahlias and when we learned that Mr. Cooper also had a small boiler he wished to dispose of, we made another purchase. The boiler was installed in the centre of our new potting shed and each morning the first arrival would stoke it up so that the shed was nice and snug by the time the potting ladies arrived.

Using Dexion, we constructed eight benches at a suitable height and the finishing touch was the building of a roller conveyor, which we ran down the side of the benches on which the ladies placed their finished boxes. These where collected from the end, placed on a trolley and pushed into the greenhouse and put on the floor where they were left to grow on. Each potting lady could pot approximately 300 plants per hour, the whole of the floor of the greenhouse would very soon be covered with batches of dahlia plants ready for sale. In those days, most of the regular staff could recognise the varieties by their foliage. These we kept together in their batches. When the appointed time came for the orders to be despatched, the potting shed became the packing shed. The ladies would stand and pack box after box of plants, wrapping them in newspaper before placing them inside our special boxes.

We had a very controlled system, which was strictly adhered to so that we could locate a specific order should the need arise, which it often did, as customers would start to get anxious about their orders. I remember one lady querying her order, saying she had only received twelve plants and not the thirteen on her order confirmation. On investigation, it was discovered that the thirteenth was 'P/P' – our abbreviation for postage and packaging.

handle carefully

When we first started packing dahlia plants we copied Ernie Cooper's method. He used recycled cardboard boxes and old jam tins supplied from the local hospital. These were used for the smaller orders of four plants, but as well as looking unprofessional, it was a very slow packing process and we found we could not keep up with the volume of orders.

In 1959 I came up with an answer by designing a corrugated cardboard box. I took it to a local company who manufactured corrugated boxes in Hatfield, and they gave me a satisfactory price The box came in two sizes, with divisions to keep the plants stable. They were a haystack shape, with a handle on the top, so that they were impossible to stack upside down. They also had special inward facing lugs that put an end to the pilfering of plants that we had experienced. To complete this design, the boxes were printed in green, with planting instructions on one side and on the other an appeal to Mr. Railwayman to 'Handle carefully.' AYLETT NURSERIES LTD appeared on both sides. This specially designed box stood us in good stead for the next 25 years.

I estimated that the extra cost of the boxes was more than outweighed by our savings in time and labour, added to by the reduction in damaged plants and the expense of subsequent replacements and by the fact that they were a good advertisement for us. After two or three seasons, we experienced problems with customers not receiving their plants within five days, which we considered the longest time that the dahlias could be shut away in a cardboard box. We discovered that the local station at St. Albans found it very difficult to handle the number of boxes that we were despatching on a Tuesday of each week.

Ralph with our specially designed dahlia boxes.
The greenhouse in the background is the reconstructed greenhouse from Marshalls Drive

After a great deal of discussion with British Rail, it was decided that the only solution was for us to take our Land Rover and trailer to London and deliver the packed dahlias directly to the different railway stations that served the various parts of the country. This meant delivering to Paddington for the West Country, Waterloo for the south, Bricklayers for Kent, Kings Cross for the east and finally Euston for the northwest.

As you can imagine, this was all very time consuming but made the delivery of dahlias much quicker. We aimed to leave the nursery at about 6 o'clock in the evening, returning home at about 10.30. Brian was the chief driver for this task, with me standing in when he couldn't make it. Thank goodness, we usually only

Dahlia arrangements staged in packing shed for our first open day

did this mammoth tour on a Tuesday, but when there was a bank holiday we had to return the following day to complete the quota. Needless to say Brian was always very thankful when June came and the dahlia orders were all despatched.

We needed to attract visitors to the nursery, so that they might view the dahlias growing and be tempted to order some plants for the next year. I looked upon this as a natural extension of us exhibiting at flower shows. So in 1959 I decided to put a small advertisement in the local paper, stating we were having an Open Day and that everyone was welcome to come along.

The dahlias were all grown in alphabetically marked beds, this made it very easy for the public to walk at their will throughout the nursery Trial Ground.

My Mother-in-law to be, Mrs. Rowland, was a keen flower arranger and offered to create six arrangements using solely dahlias, so as to demonstrate their versatility. These were displayed in the packing shed. The event was a great success and, some 500 visitors enjoyed a free cup of tea over the weekend and we vowed to repeat the event the following year.

Later that year St. Albans Round Table approached me to see if I would be willing to loan them our car and trailer, as they wanted to transport their Christmas tree. This was to

Dahlia arrangement of the variety 'Jean Fairs'

make a tour of the streets of St. Albans, playing carols through a record player which was amplified from the back of the car, with the speakers hidden under the foliage. This was to take place for the ten days before Christmas. I said I had no objections to loaning them the car and trailer, but made the stipulation that I would be the driver. I thought that this way I could keep an eye on my only means of transport. It proved great fun; I remember on one occasion the record player breaking down, obliging us to sing the carols beside the trailer as we walked through the narrow streets of St. Albans.

The Round Table raised a good deal of money for local charities during those ten days and I was so impressed with their attitude to community life that, when I was invited to become a member, I accepted with no hesitation. Through the Round Table I met a likeable chap named David Morgan. Despite the fact that both he and I had lived in St. Albans for most of our lives, our paths had never previously crossed. Through chatting to him I soon found out that he was an architect, so not unnaturally when I was in need of some planning advice, I sought out his expertise.

David as a partner of Cannon Morgan & Rheinberg has looked after our planning requirements since that day. The very first planning permission he applied for on our behalf was for the placing of a second caravan on the nursery and that was back in 1961. In the last 45 years he has made 43 planning applications, in addition to designing buildings to suit all kinds of needs as the nursery has developed. Many examples illustrate the diversity of his work, such as building over the shed that housed the electricity boards, to create a new packing shed when we outgrew the one we had originally cobbled together. At that time we also needed a larger storage area for our cut flowers during the summer. Incidentally, when it came to running our Open Days, the free cups of tea were served from this building, which also incorporated new toilets for the staff and visitors.

New storage area for cut flowers ... and cups of tea

We were able to connect these by running a salt glazed drain down through the nursery and into the sewer that the city council had recently installed in Birklands Lane. This freed us from the dreaded visits of the cleansing service, which had to be summoned whenever the septic tank was full.

During the mid 60's, we were successful in gaining planning permission for a display area for garden buildings. This was needed because a well-known manufacturer, Halls of Paddock Wood, had approached us with the idea of renting a small area of land on the front of the nursery. The money they were offering was very attractive to us, so we were delighted when we were able to welcome them on to the site. This arrangement continued for many years, until the land that they occupied became more valuable to us as part of our retail area.

The piece of land where the marquee for our open days was erected

... the venture we named 'open days' proved very popular

The venture that we had named 'Open Days' proved very popular with our regular customers and we decided to extend the event across two weekends. We also decided that instead of holding it in the packing shed, with tea being served from the office, we would hire a small marquee and erect it on a piece of land to the left of our flower shop.

In 1963 this marquee attracted some 6,000 people and 5,000 teas were served by a band of various willing helpers. These included Ada Crain and Brian's Mother, as well as Brian's girlfriend Audrey, who some four years later became his wife.

The white shed acting as the tea servery at an open day

The band of tea ladies at the ready Ada Crain to the far left and next to her Mrs. Fowler (Brian's mother)

The tea tent was attached to the main marquee and in addition to teas for the public, sausage sandwiches and apple pie were dispensed to the staff during the day to keep their spirits up as they got tired. The theme of the floral displays that year was 'The Stages of Life'. I can't remember a single one of the arrangements, but it must have been quite a feat and stretched our imagination to the full.

Over the next three years, the marquee got bigger and bigger and the number of visitors increased, as did the interest in dahlia plants. We were feeling satisfied with our full order books for the following season. The whole family were roped in to lend a hand during the days and my Auntie Ann would sit in the evenings reconciling the cash, which for her satisfaction had to balance to the penny. Auntie Ann was my Father's sister and we always looked forward to her visits. She taught Hazel the basics of balancing tills, which proved to be incredibly useful.

It wasn't all plain sailing, one year it rained all day on Friday, which made working in the marquee and picking dahlias very unpleasant. We were just finishing and looking forward to a long soak in a hot bath, when a monumental gale developed and the marquee pegs started to move. The prevailing wind was from the southwest and it looked as though the marquee was about to blow over, so we quickly moved the Land Rover into position on the grass verge of the road. We tied the guy ropes to the Land Rover's towing gear, slapped the vehicle into four-wheel drive and pulled the marquee upright. The Land Rover was left overnight in this position, which was sufficient to hold the marquee stable until the next morning, when the contractors arrived to re-stake it.

We then surveyed the damage inside. Finding that the displays had survived, but that the ground was waterlogged, we negotiated with the local farmer for some bales of straw and spread these about to soak up as much of the water as possible. Although we had to keep the marquee closed on the Saturday, by the Sunday the ground was sufficiently dry to be able to open. We hadn't foreseen that with peoples' feet stirring it up, dry straw would create a dust. It wasn't long before you couldn't see from one side of the marquee to the other and there was a good deal of coughing going on, so in haste the marquee was closed once again.

Luckily all was not lost as that was the first weekend and during the week the sun shone, returning the ground to normal. We had a very successful second weekend, but that day a decision was made that in future we would engage Piggotts, the top marquee contractors, who could offer us a better service should we experienced bad weather again.

Straw laden marquee

A bigger marquee meant more preparation work and I well remember working until 2 o'clock in the morning. I decided it was time to rethink this event so that we didn't work so hard on the Fridays, reducing ourselves to zombie-like creatures that were so tired that we were not able to care for our visitors properly on the Saturday and Sunday.

Piggotts marquee

In 1966, after talking it over with my Mother-in-law, Mrs. Rowland, who had contacts with floral arrangement societies, we decided to organise an inter-society competition, whereby the societies would do the arranging for us. We obtained a list of all the national floral arrangement societies and wrote to all those within a radius of thirty miles or thereabout. We invited them to participate and the response was very encouraging.

The attraction was that the members could come on the Friday and cut the dahlias from the trial grounds themselves and what was more, they didn't have to pay a penny for them. We did stipulate that they could incorporate other flowers, but at least seventy percent had to be our dahlias. The arrangements had to be staged in a cubicle measuring 8ft long by 5ft deep and 6ft tall. These cubicles were constructed by Hazel's Father and then lined with white muslin. The enthusiasm from the top floral art societies was tremendous, and their ingenuity was amazing. The rivalry between the groups was quite keen, but a very good spirit always prevailed and it became quite an event in their year.

Mr. Leslie Rowland — note the price of £2.95 was for a dozen plants and below Mrs. May Rowland

Enthusiasm from the top floral art societies was tremendous and their ingenuity was amazing as the above pictures show. The arrangements depicted various topics from top left Marmalade Time, Number Ten, In the Red and Olympic Hurdlers.

The individual classes with my daughter Julie in the background

At the height of this event, we had some thirty-two exhibits around the sides of the marquee. At this level of competition, judges were very important and were selected from the National Association of Flower Arrangers. Many became good friends and we looked forward to meeting them yearly and listening to their tales of various flower shows.

This event was run on the second weekend and on the first weekend we decided to run an individual flower arrangers' competition. There were six classes, any individual could enter up to three of the classes and we offered handsome cash prizes to tempt the entrants. Again, the stipulation of using seventy percent dahlias was imposed. Alongside this competition, we invited St. Albans Horticultural Society to stage their flower show at the nursery. This proved very successful and brought to the nursery a variety of horticultural subjects, greatly adding to the general interest.

These Open Day events had become very expensive to run and eventually another big decision had to be made. Should we give them up altogether? Or should we perhaps make a small entrance charge, to cover some of our expenses? After a great deal of discussion, we decided we would charge one shilling (5p) to come into the marquee.

A ticket would be issued, entitling the holder to a free cup of tea and a biscuit. We had one or two people complain, as you always do whenever a change is made, but on the whole everybody was quite happy to pay. We asked the St. Albans Horticultural Society to collect the entrance money and in return we

made a donation to their funds. This arrangement worked really well, since we didn't have to find the extra staff we would need and they were quite used to charging the public to view their shows, so no apology had to be made for the charge. It's ludicrous to think that we were so worried about charging 5p, when one considers the entrance fee charged to attend today's flower shows. Over the years we increased the 5p to 10p and eventually to 15p.

Our partnership with St. Albans Horticultural Society lasted many years until regretfully, through lack of support it was disbanded in their centenary year, it was a great loss to the city of St. Albans.

St. Albans Horticultural Society's exhibit   Collecting the 15p admission charge

In the early 60's, we had permission to park the cars on the other side of the North Orbital Road, which at that time had three lanes. A team of local special constables, who used to bring the traffic to a standstill while shepherding people safely across the road, handled it all. There was no fee for this excellent service, but you can be sure that the constables all went home carrying complimentary bunches of dahlias and were clearly as pleased as punch with them. When the road was turned into a dual carriageway, this was no longer an option and we had to rethink our parking strategy.

We managed to negotiate with the farmer who rented the field adjacent to the nursery to use it for parking the cars. He had given up growing potatoes and had turned it over to cereal crops, which were harvested by September. On one or two occasions, when the harvest was late, we held our breath that it would be cut in time. When the Electrical Apparatus Company eventually sold this field to a property developer, with the distant hope of green belt land being released for housing, the field was no longer cultivated. We were lucky in being able to continue using a corner of it for additional parking at our peak times.

We were welcoming some 6,000 people to these annual events and a question we were often asked during the year was: 'What dates are your Open Days this year, as we want to visit the nursery?' We would then explain that we were open all year round and that the dahlia Trial Ground was open to be viewed at any time during the flowering season. We realized that we'd made a marketing blunder in calling these events 'Open Days' and solved the problem by re-naming them to 'Dahlia Festivals'.

The tea tent could no longer cope with the volume of cups of tea required and we also needed the space for a larger marquee, so refreshments were moved to the back of the nursery, where the packing shed was situated. This was an area that was normally closed to the public, so we always had a mad scramble to tidy up and make the place look presentable. Our faithful team of ladies returned year after year to serve the teas and Doreen's Mother would cook the sausage sandwiches for the hard working staff.

One occasion that remains in my memory is the time that she put the baking tray, complete with the fat that she had cooked the sausages in, on the floor. She was, after all, very short of space in her makeshift kitchen. Our golden Labrador, Chips, who had a free run of the nursery and was much loved by all, found the tin. He thought it was his lucky day and polished the contents off unknown to anybody. It wasn't until he started being violently sick that it was discovered what he had eaten. Poor old Chips had to spend the night in the garage until the contents of his outraged stomach were expelled.

After the first weekend, when we'd had a local dignitary present both our prizes and the St Albans Horticultural Society awards, the marquee would be cleared out on the Sunday night. It would then stand empty, awaiting the annual return of the flower arrangers, so when the Ladies Circle approached me to see if they could use it for a charity fund raising event, I agreed without hesitation. The event evolved as a lunchtime fashion show. Some 350 guests enjoyed a three-course lunch, superbly prepared and served by the lady members.

The fashion show itself was organised by a local shop and compèred by Julia Clements, the internationally revered floral arranger. The afternoon started with Brian, dressed as a dahlia yokel, pushing a glamorous model down the walkway on decorated straw bales, adorned with dahlias in strategic places.

The fashion show was a smash hit and raised a tidy sum for the St. Albans holiday home in Worthing. This charity was dear to our hearts, as Nan Gazely, a remarkable lady who worked tirelessly for the disabled elderly of St. Albans, had founded it. When this event came to a natural conclusion, we decided to run a floral art demonstration in its place. The lunch was presented as a picnic in a box, prepared by a band of staff and handed out with a glass of wine to the guests on their arrival. When the contents of the boxes had been devoured,

Julia Clements compared the fashion show
in her usual eloquent style

One of the models

the demonstrator would take to the stage and delight the audience with an extravaganza of floral arrangements. Dahlias, of course, took centre stage.

We would pick the brains of our highly qualified judges as to who they could recommend as a demonstrator and all were agreed that the very best had to be George Smith. He could be relied on to deliver a demonstration of superb quality. George comes from Yorkshire and has that endearing quality of telling it how it is! I well remember his opening remark, 'that he had demonstrated in many places but never in a lay-by, on the M1 before!'

George Smith demonstrating 'in a lay-by'. We are still enjoying Georges wonderful demonstrations to this day – they never fail to inspire and enchant.

The Rotary Club of St. Albans Verulamium then approached me with a request to use the marquee in the evening, for a square dance. I consented, with the provision that they would guarantee to clear the marquee in preparation for us to start building our staging for the coming weekend. This was agreed and for many years a very enjoyable fun evening was attended by hundreds of people and at the same time a lot of money was raised for various charities.

The staff were all invited and although many were a little reticent at first, they all joined in and there was much hilarity as we tried to decipher what the caller was saying. A few people mixed up their lefts and rights, but nobody minded, as it was such good fun.

As the nursery developed, the piece of land where we erected the marquee was taken into the sales area, so we then had to resite it alongside the propagation house. It was there, during the 80s, that disaster nearly struck us one year. It was Sunday afternoon and the weather hadn't been very settled, but just as the competitors began arriving to dismantle their exhibits, the wind began to blow. At first it was simply unpleasant, with nothing much to worry about, but all of a sudden the sky darkened and the wind strengthened until the trees could be seen to be at a forty-five degree angle. I had just climbed the stairs to the office and looked out of the window fearing that something dreadful was about to happen.

I was horrified to see the marquee ridge drop gracefully to the ground, taking the canvas with it. There were many people inside. I called the fire brigade who arrived without delay, but luckily everybody had been evacuated by then.

... *I was horrified to see the marquee ridge drop gracefully to the ground*

Just one unfortunate lady had to be sent to hospital and only with a minor injury. Piggott's were now put to the test. After a phone call, they arrived that very evening to cart the damaged marquee away. They returned on Monday morning with a replacement, so that by Tuesday everything was back to normal and the week's festivities went ahead.

Sadly, the Dahlia Festivals came to an end in 1988. The floral arrangement societies had lost the support of those loyal bands of helpers who could invest the time and the commitment that was required to undertake the exhibits. I suppose the younger members either had very young children, or worked during the day. We didn't want to see this great event fall below the standard that we had all become accustomed to and I always believe in quitting while you are ahead. The St. Albans Horticulture Society had also disbanded and we just couldn't envisage organising the festival by ourselves. It is a great tribute to those shows, all those years ago, that I often have customers who say, 'We remember the Open Days!' We now hold a small Dahlia festival in September each year, but I have to admit that they are a mere shadow of those great shows.

... attracted a steady flow of regular customers

During the early 60s, the shop attracted a steady flow of regular customers and became a useful outlet for selling the other crops that were grown to fit in with the dahlia production. The Dutch light house, which was fondly known as the 'guinea pig' because of its origins, produced two crops of chrysanthemum blooms each year. The 'earlies' were planted in March and flowered in June, while the 'lates' were grown outside in large pots during the summer and carried into the structure to flower in October.

We sold as much as possible of the crop ourselves, or to local florists, but inevitably we had a surplus and that had to be sent to the dreaded market. Here, we had expanded our horizons, going into Birmingham and Coventry markets. To transport our flower boxes to these markets we made use of the Lea Valley carrier, Mr. Rafferty. We were the last call on his pick-up round, before he set off up the newly completed M1.

It was a lucky day when a young mum approached us for a part-time position. Her name was Ann and she took to her job like a duck to water. Ann Heward eventually moved onto the nursery, living in one of the caravans with her young daughter Susan and a number of Elkhounds. She proved invaluable in the production of our dahlias for ten years, nurturing the plants as if they were her own children. It was a sad day for the nursery when she moved with her parents to Somerset and an even sadder one when, some three years ago, we heard from Susan that Ann had died.

Another young lady that we welcomed into the business and who also lived in one of the caravans for some time, was Eileen Stamps. She too became involved with dahlia propagation, taking over Ann's job when that time came. She also had a dog, a little Jack Russell, who was quite a handful. One Saturday afternoon she was cutting the edges of her lawn when the dog, which insisted on attacking the shears, had his tongue trimmed by mistake. A distraught Eileen confronted Hazel in her garden and together they made an emergency trip to the vet to have the damaged repaired. Eileen became a good friend and would often keep the children company when we had an evening out on our own. Today, she is still involved in horticulture, having become an instructor at a college in Cheshire.

Ann Heward propagating dahlias

The total number of dahlia plants produced at that time was in excess of 40,000. Retail sales accounted for a large part of this figure and we also executed some wholesale orders to various cut flower growers or parks departments. This was the era of the horticultural mail order business, when the public pored over glossy catalogues before making their choice.

We were approached by one firm that sold directly to the public by means of their colour brochure. They were not growers, but merely a packing house and consequently they didn't know one plant from another. They sold at exceptionally cheap prices. We thought this was an opportunity to clear any surplus dahlias, earning ourselves a few pounds into the bargain and although all the dahlias were labelled, I have to put my hand up and say that many of them were not true to form. We reckoned that people who bought from such a cheapjack firm would never know the difference between one red cactus dahlia and another. A red dahlia is a red dahlia. In any event, we never had a complaint and we cleared our crop and earned ourselves some much needed revenue. As soon as we could, we gave up this rather unsavoury practice.

Now that we had our half-acre of heated glass, we were desperate to find crops that would fit in with the dahlias and keep the cash flow healthy. I well remember growing a beautiful crop of pot hyacinths, which I thought we would have no trouble in selling. As luck would have it, there was over-production of hyacinths that year and all ours ended up on the rubbish heap.

As I had made a deal with the supplier of the bulbs that I would pay for them when they matured, I was somewhat embarrassed. I am perfectly sure that if that had happened today, we would have been put into receivership, but as it was then, there was very little the supplier could do, especially as he was in Holland. The bill was settled in installments and the debt was eventually cleared, but it was not a happy experience.

We also tried growing some early forced tulips and daffodils, with the idea of selling them through the shop. Although these were successful, they came very quickly to maturity and we couldn't sell the whole crop through the shop. Again we had to use the markets, with little financial joy. We did, however, persevere with this project for a number of years.

... and all ours ended up on the rubbish heap!

Although I say it myself, we grew the most amazing crop of 10-week stocks. The seedlings were transplanted directly into the ground in the English Light greenhouse, with the light green seedlings being discarded, as these would be the unattractive single ones. The scent was absolutely wonderful as they came into flower, and I well remember spending the Whitsun bank holiday with Hazel, picking and packing them for market.

Another crop we tried was pot cinerarias. We sowed the seed in the propagation house during August, pricking out the seedlings and finally potting them into 5½in pots, which were put onto the floor of the English Light greenhouse, and grown on to mature during March and April, when the space was needed for the dahlias. They were magnificent and not at all like the pathetic specimens we are offered today. This is probably because they were so labour intensive. The watering had to be carried out by hand and by an experienced person, and they were also expensive to market, being so large so that you could not get many to a market trolley. I remember delivering to a market wholesaler and the plants would not fit on the lift so I had to walk them up the stairs. After we had shut the shop and had a bite to eat, we would set about selecting the plants for the following day's market. We would work until the light faded and we couldn't see the colours. Although we got tired, we all enjoyed the task and would work as a happy team, always with the hope that the following day's market would return satisfactory prices. Unfortunately, this was rarely the case.

There was a demand for summer bedding plants and we managed to fit into our busy schedule the growing of about 2,000 boxes of them. The production of bedding was standardised throughout the country, all plants being grown in wooden boxes that held 8 x 5 rows of plants, making 40 in total. That was much easier than the variations that we have today, with so many combinations available.

The selection of types was not immense and consisted mainly of lobelia, alyssum, antirrhinum, salvias, stocks and asters. But then the planting schemes were not as ambitious as they are today. If one of our customers wanted just half a box, that gave us no problem. We just cut the box in two with a sharp knife – it was all very easy.

After the English light house had finished with the production of dahlia plants, it was rotovated and prepared for the next crop, namely some 20,000 chrysanthemums for cut flower. These varieties flowered naturally from early October

through to the end of December, they were marketed in exactly the same way as the other crops and experienced exactly the same difficulties with the markets

Hazel holding white Chrysanthemums

We had given up on runner beans, having declared them a monumental disaster, but the bean posts didn't go to waste. I devised a plan, using the 'guinea pig' greenhouse as a guide line and built a Dutch light structure in front of the packing shed. The bean posts were used as uprights and the cross members were made from the sides of the lettuce frames that we had cut in half. Finally the roof was glazed using the lights: the production of lettuce was transferred into the English light greenhouse, as being heated, it could produce an earlier crop.
The only purchase that had to be made was the wooden gutters. This greenhouse was then used to harden the dahlias off and after that an Italian worker, named Jo, planted and cultivated a crop of tomatoes in true Italian style. The structure remained in production until the late 70s, so we can't have made such a bad job of it.

The outside land was mainly put down to dahlia production. I haven't mentioned the raspberries we had first planted, way back in that very first year. Although these were one of the better crops we grew, they were not as successful since they had such a limited shelf life. If the market didn't sell them the very first day they arrived, they were thrown out. So these were scrubbed, but a few canes escaped our boundaries and can still be seen today at the bottom of the adjacent field. Along with the raspberries, we had quite a few Boskoop Giant blackcurrant bushes planted on the poor land behind the garden shop. These fruited at a time when we were busy with the dahlias, so we advertised for pickers to be paid at piece rate, which is based on the output of fruit picked.

The patients from the local psychiatric hospital across the road applied as pickers, and seeing no objection to this, we agreed that, as long as their nurse was in attendance, they could work on the same terms as anyone else. They duly arrived, but proved a little unsatisfactory in that they ate more than they picked.
They were observed throwing the fruit into the air and catching the berries in their mouths. I'm sure they enjoyed themselves and that it gave them a change from their normal routine, but it wasn't financially viable for us. As a result of this venture, one of the pickers asked if he could have a job with us. This gentleman was with us for many summers, he would arrive to do the hoeing and other simple tasks. He was a lovely man, well educated and highly intelligent. He had studied to be a Roman Catholic Priest, but something had gone wrong and he had overtaxed his brain. In his work with us he found a little peace of mind, but you would often find him in floods of tears, saying the Rosary prayers.

The growing of all these different crops was fitted around a busy dahlia schedule. We were continuing with our exhibiting and at the same time trying hard to build up the floristry business. Both of these commitments were very demanding and we had very little time to ourselves, barely enough for a few days holiday each year.

Friday afternoon was wages time and I was the wages clerk. It makes me shudder when I think of the scant records that I kept, especially when compared to today's vast files. My records were all kept in a red accounting book, in which I listed the names of employees week by week and calculated their pay, making the standard deduction for the National Insurance stamp. In those days, every employee had a National Insurance card, which they presented when they first started work. If someone left your employment, they collected their stamped card from you, ready to present to their next employer. This was before the introduction of the graduated pension deduction, the stamps differed for married women and men, but did not vary week to week, as long as there had been no increase in the countrys budget. The stamps were bought over the counter at any Post Office and when the card was full it was the employer's responsibility to change it for a new one. This was a long, drawn out process that involved attending the local N.I. office and sitting on very hard seats until it was your turn to approach the desk. It could take hours. It was the unfriendliest of places, with little or no privacy. The person behind the desk would then transfer by hand all the details from the old card to the new card. Whenever I remember this ordeal, I have to admit that the present day system is a dream by comparison.

Bookwork was a chore that I absolutely hated. After all, it was keeping me away from the important jobs in my life and I was very remiss in ruling off the bottom line. Consequently, when the time came to change cards, or at a year-end, it was sheer hell to balance the books. One Friday I was sitting at my desk, struggling with this task, when Hawker Siddley decided to test their latest aircraft engine in the wind tunnel just up the road. Not being in the best of humours, I picked up the phone and very firmly told them to cut the noise level down, as I couldn't concentrate and I was trying to complete the wages. It was soon after this that Hazel came to the rescue and took over the job.

As a result of the nursery's specialisation of dahlias and our exhibits, Roy Ward of Wards of Sarratt, who grew and exhibited dahlias on a similar scale to us, approached me to ask if I would be interested in joining the British Dahlias Growers Association. This was an association formed by dahlia growers, with the idea of getting together in their local regions to discuss the many facets of dahlia growing. The local branch met twice a year and the annual general meeting took place in October, with a grand get-together of the countrywide membership.

I attended my first regional meeting at Agriculture House in Knightsbridge, London. There were a number of growers that I had met at shows and they made me very welcome. I remember discussing at some length the retail price that we would charge the following season, which ranged from two shillings and sixpence to three shillings. At that time there was no shortage of dahlia growers and at the first R.H.S. Great Autumn Show we attended, I counted thirteen other stands showing the one genus – the dahlia.

Hazel and I went to our first AGM of the association just after we had become engaged. It was held in Harrogate and neither of us had ever been there before. All the members took along their ladies; it was treated as a little holiday after the vigour of the busy season. Dahlia growers tended not to have proper summer holidays. While the gentlemen attended their meeting, the ladies, having been given strict budgets by their men folk, would take advantage of the local shops, stopping off to take coffee together and to catch up with what had happened in their lives during the year. Not all the ladies worked with their husbands; in fact very few did and consequently had little interest in dahlia growing. Hazel was the exception and always wanted to know what went on in the meetings.

We met some really fascinating people and they took us youngsters to their hearts, showing great interest in what we were doing back in St. Albans. Dahlias were very popular in the north of England and the growers were great characters and the kindest of people. They had great warmth and friendliness and they certainly knew how to make us welcome.

It was at Harrogate that we met Mr. and Mrs. Jackson of Birkenhead, who grew dahlias as cut flowers and made a success of it. Their cut dahlias were superb, and when they invited us to their nursery, we jumped at the chance. It was a visit that encouraged and inspired us. We came away having made two good friends, as well as acquiring a few tips about growing dahlias.

We looked forward to the annual get-together, meeting in Edinburgh, at Kendal in the Lake District and at Pithlochrie. We always made a point of staying away for a few days longer than the meetings and would go and visit our grower friends at their nurseries. We never came away without a few good ideas for the following season. At one regional meeting there was a great deal of discussion regarding control of virus in dahlias and it was suggested we should consult Rothamsted experimental station in Hertfordshire to see if they could help. At the next meeting, Dr. R. Cammack, who was their senior scientific officer, addressed the members. He suggested we should grow our dahlias under mercury phosphorous lights, to achieve stronger growth and to increase the productivity from the tubers.

As the most local member, I was asked if I would help with a trial. I agreed and we gave the station twelve of our tubers to grow under their controlled conditions using lights. We did the same, using similar tubers but under our normal conditions.

Five weeks later, we had produced fifteen cuttings while the station had produced fifty. Realising that our bench space could be reduced, we installed one mercury phosphorous lamp. Our results were very good, but not as good as the station's. They grew at a much higher temperature than we did. The other advantage of using this light was that the cuttings rooted much faster.

Dr Cammack's next suggestion was to convert a greenhouse to be aphid-proof, as it was found that the aphid spread the viruses from plant to plant. We had to remove the ventilators and install a high-speed fan to suck the air into the house, thereby pressurising the greenhouse. This part of the experiment was not as successful as the lighting project. It took us three seasons to achieve our virus-free tuber from which we could propagate. At the same time it was suggested that we should dip our knives into tri-sodium orthophosphate to sterilise them between each cutting. This process certainly slowed us down, but we religiously adhered to it.

Mercury Phosphorous lamp

The following season, in 1966, it was our turn to host the meeting and included in the agenda was a visit to the Lea Valley Experimental Station, which had also been carrying out trials on behalf of the British dahlia growers. They had experimented with growing plants in small pots, in an insect-proof glasshouse, giving them short day treatment to produce pot tubers. These were used to propagate from the following spring and would supposedly be free of virus.

After the work with Rothamsted and Lea Valley I was asked by the Principal of Oaklands to present a paper entitled 'At the flick of a switch'. I shared our lighting experiences with an assembled audience of growers and students. We unquestionably learnt a lot about the viruses that are present in dahlias, but the work never came to a satisfactory conclusion. Whether this was due to a lack of funds, or the fact that the scientists deemed the problem insurmountable, is anybody's guess. In any event, the experiments were discontinued and we growers learnt to live with viruses, applying very vigorous rouging to keep them under control. The most useful lesson that we learnt was about the lighting. Installing more lights enabled us to propagate successfully some the varieties that were shy of producing cuttings. We were thus able to increase the output of rooted cuttings without expansion of the glasshouse area.

... the stage was set for me to move forward

At the beginning of 1962, the stage was set for me to move forward and achieve my goal of running a successful business. To the outside world, I had already achieved that by appearing to be flourishing and by winning gold medal after gold medal. Only the family and the bank manager knew the truth! I had built the glass I needed and I had a little capital in reserve to take care of the unknown, so what could possibly go wrong now?

Feeling more confident in my future, I decided that my three-year engagement to Hazel was long enough and set the date for our wedding. This was to be in January, so as not to interfere with the growing year. We decided to rent a flat in Seymour Road, about three miles away from the nursery. We found this a little difficult, especially if we had problems with the boiler at night and Ralph had to call me out to put them right. We also detested living in a flat. Being without a garden was something that we both hated and after about six months we saw a little bungalow being built, just half a mile away from the nursery. As our rental contract was about to expire, we made enquires of the estate agent. I can't remember how much it was, but once again our dear friend Mr. Peasley came to the rescue, personally loaning us the necessary funds to be able to purchase the property.

This was a great improvement as we took little time in going backwards and forwards, which made boiler duty a little easier. Luckily for me, we had a very kind neighbour who lived adjacent to the nursery. He was always phoning up late at night, saying the boiler alarm was going off. This was our only means of knowing when the boilers were malfunctioning. If the weather was very cold, the only way you could check on them was by physically making a visit.

The winter of 1962 was exceptionally harsh and to keep the boilers up to temperature we had to stoke them every three hours, including all through the night. Ralph and I took it in turns, alternating our shifts so that at least one of us could get a decent night's sleep. The boilers were overworked and at the end of the winter various leaks started appearing in the tubes. They had simply come to the end of their useful life and I had to think of an alternative. Again, our friend Rafferty put me in touch with a company near Staines that specialized in second hand boilers. After consultation with the owner, I purchased two oil-fired sectional boilers. This decision was taken because at that time oil was cheap and we wouldn't have to stoke them physically at night which, after the kind of winter we had just experienced, seemed a good idea.

The old sectional coal fired boiler being removed

First, we had to dismantle the old boilers. This wasn't too hard, as they practically fell to bits. Money was still rather tight and again we could not afford to employ outside contractors to install the new boilers. Ralph and I, together with other members of staff, went about completing the remedial work that was necessary before we could lower the sixteen sections that went to make up one boiler down onto the plinths in the boiler house. Three tapered rings, one at the top and one at each side at the bottom held each section together. As each new section was added, you had to pull the section up tight to the next section, so the two boilers took a considerable length of time to install. We certainly learnt an awful lot but heating engineers we were not.

The next stage was certainly an eye-opener, as we had to install two tanks, which had capacities of 1,300 and 1,500 gallons of oil. Next, it was time to install the oil burners and the electrics and for this we engaged experts, as we thought these jobs were beyond our capabilities. The work was finished by the summer and we looked forward to a trouble free winter as far as the heating was concerned.

My Father retired from L. Rose and Co. Ltd. in December 1962, after thirty-three years as company secretary. He was looking forward to his retirement and to having more time to devote to Aylett Nurseries. My parents decided to sell the house in Mile House Lane, which was much too large for just two people and to build a bungalow in a plot of land that they created by dividing the large garden in two. This way they could remain close to us, to the nursery, and to their dear friends, the Peasleys.

The housing market at that time was none too buoyant and Father wanted the sale of the big house to be finalized before he started on the building of the new bungalow. So it was decided that they would move in with us at Mile House Close, as an interim measure. Sadly, tragedy struck with the death of Mr. Peasley. He had been unwell for a little while, but his death was a great shock to us all. Then Father became unwell and although the doctor treating him raised no great fears for his general heath, he did not seem to improve.

In the September 1963 he also passed away, leaving a great void in my life as I had lost not only my Father, but also his vast knowledge and experience as a businessman. Life had to go on, as he would have wished. Hazel took a team and went to Thame Show the very next day. She always remembers this as being the very worst show she attended.

Mother had her new home completed and moved in, but she missed Father terribly and was never really happy. The nursery was in crisis. After nine years' trading we still had not managed to make a profit and the bank manager was losing patience with us. After perusing our annual accounts in 1964, he advised my Mother not to invest any more of her capital and recommended early closure of the nursery, before we lost any more money. Luckily she did not heed this advice and we continued to do our best to achieve financial success.

We carried on with our busy show schedule and early one October morning, when we were arranging dahlias in a local delicatessen in the middle of St. Albans, Hazel felt decidedly unwell. A visit to the doctor confirmed that she was expecting our first child and a quick count on our fingers made us realize that the baby would arrive the following May. This was not ideal timing, but we felt sure we would manage.

Mother was very lonely and we heard of a litter of golden Labrador puppies and decided to buy one for her. She had a great love of dogs and we felt that it might cheer her up. She was however, unbeknown to us, very unwell. Hazel, on the contrary, kept incredibly well and continued working throughout her pregnancy. The middle of May came and the baby was due, but obligingly decided to wait until after the peak time of year, arriving on the 2nd June 1965. Julie was a perfect baby and was no trouble at all, but Hazel took a little while to recover. Ada Crain came to the rescue and offered to act as housekeeper and nanny. She never went back to the nursery to work and instead looked after the family for many years.

The fact that Mother was unwell marred the joy of our having a daughter and she was admitted into St Thomas's Hospital in London. It was soon discovered that she had advanced lung cancer. She slipped into a coma shortly afterwards and two years after my Father's death I found myself having to cope with my Mother's, too.

We now inherited a dog named Mr. Chips. He was only a year old and quite a handful. Hazel was not used to having a dog and she found it hard to cope with both a pet and a new baby. I remember Chips taking the washing off the line, and redistributing it around the garden, which did nothing to endear him to her. I decided that he had better accompany me to the nursery on a daily basis. He became loved by all, a great character who was always first to arrive when the tea bell was rung, hopeful of the odd biscuit. He could recognise friendly vans that arrived at the gate and when one Mr. Kendal arrived to collect his flowers, Chips would be first on the scene to welcome him. He was always rewarded with a biscuit, of course.

We did not have an afternoon tea break and we realized one day that Chips always walked up the road at the same time in the afternoon; the road was then nothing like as busy as it is today. We decided to follow him and found that he went to the next nursery to join them for their tea. Yes, he was certainly quite a character. He was so well known that we featured him on the front page of our catalogue in 1968 and he even had a dahlia named after him. In his later life he became quite content to stay at home and by this time Hazel had also fallen under his spell and the children adored him. We have never been without a Labrador since, but Mr. Chips has a special place in our memories.

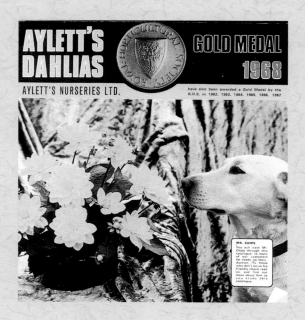

My Mother's untimely death had helped my financial standing, and although I would most certainly have wished it any other way, it seemed to be a turning point in my life. The first momentous decision I made was to build a bungalow on the nursery, enabling me to be accessible at all times and also to be on hand for my growing family.

Hazel had decided that an only child was not for her and we were expecting our second baby in July 1967. We purchased from the nursery a poor piece of land, in the corner of the site adjacent to the shop, which could only just manage to grow blackcurrants and was not important. We learned that this was where the cowsheds had stood many years previously, when the site had been part of a farm.

To save money, we decided that we would do without any architect's fees and do the designing ourselves. This worked very well until the builder went broke,

Our 1968 catalogue featuring Mr. Chips and his dahlia variety and we were left with the shell of a bungalow and very little else. Fortunately, David Morgan came to the rescue and after assessing what work needed completing, found us another builder who was willing to take over.

Eventually we moved into our new home, somewhat later than we had planned. Sally, our second child, was born just one week later. We have never regretted our decision to move onto the nursery. People often say, 'But you're never away from the job!' That's true, of course, but as a direct result I never missed out on the children's progress and they grew up accepting that Daddy was over in the nursery and always knew where to find me if they wanted to.

We then decided to approach the Agricultural Credit Corporation, in London, to see if they would finance the building of a new block of glass, to be sited between the shop and the original office. This block of glass would measure 100ft by 80ft, designed by Robinsons of Winchester and constructed of aluminium. I believed it would allow us to force dahlias and to exhibit at Chelsea once more.

It was fortuitous that the person who interviewed me at this organisation was an old acquaintance from my Oaklands days. Who says that the old school tie doesn't open doors?! After a great deal of discussion, I was granted a £3,500 loan. Robinsons were to supply the materials, but unfortunately we had not been granted enough to have the glasshouse erected and after some lengthy negotiations with Robinsons, I decided that we would assemble the glasshouse ourselves. It was agreed that their very senior surveyor would be provided for one day, to mark the position of the uprights.

Mr. Maynard actually stayed with us for two days, until he was satisfied that we were well underway with bolting the uprights to the gutter and trusses. The entire success of the operation was totally dependant on our achieving the correct one-in-ten fall, otherwise the rainwater would not drain from the glasshouse roof and would cause us a problem.

The whole framework had to be square and took a long time to construct. It began to look like a giant Meccano set as the purling rail and the ridge were added. Total precision was called for and Brian Fowler and I spent a lot of time making sure that the structure was correctly placed. The job of putting the glass into the glazing bars really taxed our endurance. If I tell you that on the first day of glazing, Brian and I broke more glass than we actually placed in the roof,

The newly constructed Robinson greenhouse beside the old shop. The dahlias in the foreground had been frosted, hence their dismal appearance

chapter 9

total precision

you will get an idea of the enormity of the task. That night we decided that we would have to get somebody else to help us fit the glass if we couldn't do better and I hardly slept a wink. The next morning I went back to try once more and to my delight the job didn't seem so bad. By lunchtime that day, we had glazed half of one of the four houses and from then on we made good progress. While we glazed the sides, our two faithful Italians, Jo and Giuseppe, laid the concrete blocks around the perimeter. When it was all completed, we stood back and admired our handiwork. Although neither Brian nor I had any intentions of going into the glasshouse construction business, we felt we had achieved a lot.
We must have done a good job, as that greenhouse is still there today full of our houseplants, having successfully weathered the hurricane of the 80s.

The house was finished, but not in time to grow the dahlias to be ready for Chelsea. It was late April and our customers were arriving to collect their plants, so rather than leave our greenhouse empty, we filled the first house with geraniums and fuchsias for sale. This was a great success and during the summer, in between all the other jobs of growing the dahlias and going to flower shows, we managed to heat this brand-new greenhouse and were able to fill one house with houseplants at Christmas time.

As a result of this, the flow of customers began to rise. They seemed to feel that here was one big nursery, rather than two small nurseries, which must have been the impression we gave before. It must have looked as if the garden shop was a separate enterprise, with somebody else's nursery behind it.

Suddenly we were in business and the gardening public wanted to buy our produce. We never went back to Chelsea, as we were always too busy looking after our customers at that peak time in May, so perhaps that guardian angel was looking after me again, for if the greenhouse had been finished in time dahlias would have certainly been the crop that I'd have been growing.

So as I said, the building of this greenhouse was a real turning point for the business and by 1970 we had managed to make a profit. Sadly, neither of my parents were alive to witness this event. I'm sure that if they had been, they would have been delighted and I, for my part, would have loved to say 'Thank you for your belief in me.' But there you are: there are some things you just cannot arrange.

It was in the 60s that I first experienced the world of entertainment. The filming of Graham Greene's well known novel 'Travels with my Aunt' brought about our first venture into the film world. This was to be shot in spring. Dahlias play an important part in the story, for it is in the dahlia border that the urn containing Henry's Mothers ashes stands. Somebody must have told the filmmakers how they could overcome the fact that the dahlia flowers in the early autumn, and not in spring. They commissioned us to provide dahlias in flower, as had been seen at the Chelsea Flower Show. The timescale was exactly the same as the flowers we had grown for Chelsea, so we could see no problems and the financial rewards were, to say the least, lucrative. The dahlias in flower were duly delivered to the film set and we were assured that when production was completed, we would receive an invitation to see the finished film. They were true to their word and the

following autumn we were delighted to receive an invitation to the premiere in London's Leicester Square. Maggie Smith had the lead part of Aunt Augusta and it was a great thrill to rub shoulders with the rich and famous. The dahlias appeared right at the beginning of the film and Hazel and I were on the edge of our seats waiting to see their debut. Much to our horror, as the camera panned around the dahlia border, a laburnum tree in full flower graced the screen. We squirmed in our seats, but I very much doubt if anybody else noticed. It certainly didn't diminish our enjoyment of a very momentous evening. Dahlias don't feature in many films, so our brush with stardom was short-lived. We did supply the broomsticks for the Harry Potter films, but sadly no invitation to the premiere accompanied that transaction.

... an invitation to the premiere in London's Leicester Square

During the summer of 1972, I had a phone call from Anglia Television, inquiring if they could bring their gardening team, led by Geoff Hamilton, to make a film about dahlias. We were always glad of any free publicity and welcomed them with open arms. On the appointed day the team arrived and we were somewhat bemused by the quantity of personnel they brought with them – a whole minibus-full. I am sure they were all essential for the production, but it seemed to us hardworking people that they spent a large amount of their time lounging about in the sunshine enjoying the fresh air.

The project took two days to complete and consumed 2,500ft of film. We thought 'Wow! This is going to be something quite spectacular!' I had to do a small interview with Geoff in the middle of the dahlia field and for various reasons endless takes were necessary. One near-perfect shot had to be redone, as a tractor rumbled by on the road. The result finished off my television career before it had

1973 Anglia Television Minibus   Cameras about to roll

even started. We were all very keen to view the nursery's debut on television, but there was one small problem: we couldn't actually receive Anglia on our television sets. We found out that nearby Hatfield was able to receive the signal, so I hired a room at the Comet Hotel and invited the staff to join us to view the programme. I remember that it was shown at 9 o'clock in the evening, which was after Julie and Sally's normal bedtime, but as a special treat they were allowed to accompany us. After some refreshment, we all seated ourselves around an especially large TV set. The film lasted just 12 minutes and was not the spectacular production we were all expecting. To crown it all, the children fell asleep and missed their Father making his first appearance on television.

The original presenter of the Gardener's World programme was none other than Percy Thrower, of Shrewsbury Flower Show fame, lovingly known as 'the nation's head gardener'. A certain Arthur Billitt, who had retired from Boots the chemist, where he had been a director of their research centre, had been asked to develop part of his abundant land into a purpose-built garden. The soil at Clacks Farm in Worcestershire was sublime and a pure joy to work with. The cables for the filming were permanently installed within the borders of the paths and the whole site had been laid out as a film set, although very unobtrusively. We had supplied the plants for the creation of a dahlia border and in return had been invited to witness the making of a programme. We often watched Gardener's World, partly because we were great fans of Percy's, but also to see which products or new plants he would be talking about. I could bank on the fact that the very next day in the nursery, I would be asked for them by his loyal devotees. We had noticed that on occasions Percy seemed short of breath and we had remarked on this to each other, being concerned for his health. Hazel and I duly arrived at the farm sometime in the morning, for the invitation had included lunch. The filming during the morning was fascinating and it was a real insight into just how much preparation work had to be done before a single shot could be taken. When everybody was satisfied that they had enough film to produce a complete programme after cutting and editing, it was time for lunch. Arthur's wife was an excellent cook and hostess and the lunch was not only delicious but also very ample. After a thoroughly enjoyable meal, we realized why we had been struck by Percy's shortness of breath in some of the programmes. Following a quick change of garments, back he went into the garden to shoot the following week's edition. Following that memorable visit, whenever we watched the programme, we would play our 'spot the before and after lunch' game, no longer concerned for Percy's health.

My next television appearance is just a blur in my memory, but I know I travelled to the studio at Pebble Mill in Birmingham to talk to Peter Seabrook on his lunchtime gardening programme. Hazel and the children accompanied me, as did a quantity of cut dahlias, of a variety that we grew at that time named 'Television'. It was a horrible orange small cactus variety, with thin incurving petals, but just because of its name it was extremely popular with the public. Hazel and the children sat in the audience as I said my bit, which must have been unremarkable because they simply can't remember the day at all.

... the early 1970's brought the winter of discontent

The early 70's was not a happy time for the country as a whole and has become known as the 'winter of discontent', with a three-day working week for many people. There were power cuts and fuel shortages, with the result that we were never quite sure how we could keep the temperature high enough for the survival of our valuable stock plants.

We were luckier than most, as our electricity supply was on the same circuit as Napsbury Hospital, which had some degree of priority, since darkness would upset the patients. But even so, we had some very anxious moments and I collected a few grey hairs. I decided that I would never be put in this terrible position again and during the next summer installed a generator for use in emergencies. It was capable of producing 50Kva of electricity and I envisaged that this would be enough power to keep the tills operating, as well as producing sufficient light to ensure that our customers were not inconvenienced. The next winter, although the terrible power cuts did not return, I was able to sleep better in the knowledge that we were secure, at least with the electricity supply.

During this time the business expanded very quickly. We outgrew the basic buildings, the first major extension we asked David Morgan to design was a glass canopy in front of the existing shop, this was to create space for the stocking of garden products which linked to our plants. The planning permission was given and the structure was built. At the same time, we partitioned off a piece of land behind the shop, where we'd grown dahlias, to create a small display area for trees, shrubs and herbaceous plants.

Glass fronted extension. The original shop roof can be seen behind the fascia board

The display area behind the shop

These plants sold readily, so much so that we soon needed another cash desk to cope with the flow of customers. We quickly realized that there was a definite demand for quality plants and it was much more profitable to sell direct to the retail customer. So during the next three years we turned again to our trusty architect, this time to create: toilet facilities for our visitors with a store behind, better office space and a workroom for our flower shop.

The sales of our garden products grew at a very fast rate. We soon had to extend the glass canopy along the front, until it joined up with the store and toilet building, to accommodate another cash register. The area we had allocated to the outside plants very quickly became inadequate and had to be expanded, at the same time we created a covered area for bags of compost and other heavy items. We never envisaged in our wildest dreams that these improvements would not be sufficient to last us for many years to come. We now had a balance sheet that was looking a little healthier and we found that the new bank manager was keen to see the business develop. He even suggested that we didn't need the Agricultural Credit Corporation loan, as the bank was quite happy to take over that debt, furthermore he talked about extra borrowing – what a change in the bank's thinking!

All these extensions we made, while serving their purposes admirably, were really just extensions to what was the very basic shed-like building in the centre. The floor was at different levels and it was not unlike a rabbit warren. When David Morgan suggested that we could construct a purpose-built, steel frame structure over the top of all these buildings, whilst continuing with normal trading, we thought it a splendid idea. We went to the Bank Manager who had no qualms about lending us the necessary funds.

This building enabled us to demolish the original shop and its extensions, leaving us with a large single building that housed our floristry department, together with our garden shop. A one-way system was introduced so that all customers were directed through a corridor into the outdoor planteria, before entering either the houseplant greenhouse, or the flower shop. This tempted them to add to their shopping baskets before arriving at the tills.

The interior of the new steel framed shop   First shops demolition

Upstairs office
View of the newly constructed planteria
Back of first shop
Entrance to the garden centre
1984 rebuild of shop

... this was the beginning of the garden centre that is so familiar to us today

Having built up a reputation for excellence, the flower shop was now thriving. David managed to persuade the planning department to let us build an office on the first floor, thereby releasing our ground floor office to extend the flower shop.

The business went from strength to strength and we soon had to make still more changes to our planteria. We took a piece of the dahlia field and turned it into a car park. The grass that was used for the Open Day marquee at the front of the nursery was bulldozed away, to make room for a new roadway that replaced the one that had been so carefully constructed in 1955. This new roadway now ran alongside the boundary fence. The new area created by this exercise was fenced off and our planteria was created with sales beds for plants and interconnecting paths for easy customer access.

This was the beginning of the garden centre that is so familiar to us today. It was not at all like the nursery that I had first envisaged, but I had grown older and wiser, I now knew that growing crops in small quantities for the wholesale market was a very unwise and uncommercial undertaking and that $7\frac{1}{2}$ acres is not the ideal size for a wholesale nursery. It conveys some idea of how quickly the business developed when I tell you that in just seven years the wonderful new shop, which I had envisaged would last us forever, was proving to be inadequate. This was a period of enormous expansion of garden centres throughout the whole country, as everyone realized that the general public preferred this way of making purchases for their gardens. They seemed to prefer it to studying catalogues, never quite knowing what they were buying, or to trudging around flower shows.

Architect David again came up with a stunning plan for a new shop, doubling the space. I don't think I could have come to terms with the demolition of the steel structure so soon after its completion, if I hadn't had the assurance of the experts that it could be erected elsewhere on the site. The thought in the back of my mind was to use it as a store, so that in the winter, when we had time on our hands, we could fill it with backup stock. This would help us in the spring, when our suppliers were inclined to run short and we were pressed for time. Wholesalers were happy to offer advantageous terms, too, as it saved them space in their warehouses.

The whole building was gutted, leaving only the two brick ends which housed the toilets and store on one side and the office and flower shop on the other. A steel framework, some 60ft wide, was erected. This became our spacious new shop and was equipped with new shop fittings, to display a wider range of garden sundries. There was now room for four cash registers, which were situated by the exit. Our customers were able to shop in our planteria and then come inside for a tempting range of china, silk flowers and garden sundries. At Christmas time we built a special display to promote seasonal items. The new retail area met with the approval of our loyal customers and continued with its healthy growth. Three years later I had the satisfaction of seeing the 'old shop' erected at the back of the nursery to act as our new store joining it on to the existing one (see page 81). Both buildings are still in existence, although both have been outgrown yet again and extended.

Fifteen years had begun to take their toll on our 'guinea pig' Dutch light house. There were grants to be had for demolishing old wooden glasshouses and replacing them with new aluminium houses and we were only too happy to take advantage of any government handouts. An added advantage of these new glasshouses was that the ventilation was automated, whereas the old vents had to be operated by hand. This was a vital task, but one that took a vast amount of time out of the working day. In the summer, when the vents could not be closed until the cool of the evening, after the staff had finished, it was another task I undertook myself.

The cropping plan that I had in mind for this new greenhouse was geraniums for the spring retail trade, followed by a crop of chrysanthemums, to be sold as cut flowers, either through our own flower shop or to local florists. Hazel's Father had recently retired and being the sort of person who liked to be busy, joined us on a part-time basis. When not applying his skills in constructing almost anything in wood, he served as a delivery 'boy'. He was yet another multi-tasking member of the family. So when we required some frames to be placed on top of concrete blocks to act as temporary benching to grow the geranium crop on, he set to work and soon had an enormous quantity ready. He never did anything by halves and today those frames are still in use, no longer for the job that he made them for, but for keeping the rabbits out of our Dahlia Trial Ground at Willows Farm.

The English light greenhouses, which we had worked so hard to dismantle at Uxbridge, had proved a great success in those early days but now needed a great deal of maintenance. The light factor they provided was not anywhere as good as the new type of aluminium glasshouse. This time, we had funds to have the greenhouse erected by professionals and the project was undertaken in two phases, over a period of three years. At the same time I decided we should change the 4in. cast iron heating pipes and installed the 1¼ in. piping that are now considered an industry standard. We completed this work ourselves, leaving the construction of the glasshouse to Robinsons' staff.

I had read a lot about thermal screens. Up until this time, the heat loss from the roofs of glasshouses was something we all accepted. But the government was now offering healthy grants for improvements that would save the world's diminishing fossil fuels, thermal screens were said to fit the bill, saving at least twenty percent of heat loss through the glasshouse roof. So when our new glass was finished, we had a specialist company install thermal screens. They were operated electrically, on a time clock. With a flick of a switch in the evening, the screens opened to block out the top half of the greenhouse. They proved to be a sound investment and probably surpassed the savings that had been claimed by the Ministry for them. An added advantage was that in the heat of summer they could also be used as very effective shading. Today, they are an essential part of any glasshouse development, having been improved mechanically and the material upgraded to aluminium foil.

On one of our annual pilgrimages to the British Growers Look Ahead exhibition (B.G.L.A.), we were fascinated to visit a stand demonstrating a potting machine. We were still potting by hand and it was becoming more and more difficult to recruit part-time ladies to keep up with our vast demands.

We took the plunge and purchased a brand-new potting machine from Germany. This machine could pot 1,200 dahlia plants an hour, with just two people to keep it supplied with potting material. It was a wonderful improvement and meant that staff could be released for other duties. It was at this same exhibition that we found a machine to improve our soil mixing procedure. It seemed a far cry from the time when all our potting soil had to be mixed by hand and barrowed away. This new machine just needed to be fed with peat, sand, perlite and the required fertilizer and – hey presto! – perfect potting soil.

The two caravans that were really mobile homes had served us well over the years, providing accommodation for a succession of staff members. Our trusty

architect advised us that if we agreed to give up the planning permission we had for the 2-caravan site, we might persuade the planning authorities to grant us permission to build a brick building. This was to be used as a much – needed store, replacing our old packing shed, with a second storey that would provide accommodation for staff. Luckily, the district council approved and the plans were drawn up.

Brian using the new potting machine

We were finding that it was difficult to recruit students that wanted to complete their practical year with us before embarking on their courses at Oaklands. Finding lodgings in St. Albans was near impossible. The new building had three individual units suitable for student accommodation, as well as a small flat which we envisaged would be ideal for a more senior member of staff, who could relieve me of some of my evening duties.

The building was completed and it certainly was a great advantage to be able to offer accommodation to our students. A few years later, a young man who had been with us since leaving school and showed great promise as a budding horticulturist, moved into the flat with his bride. That young man was Tony Day. He is now our production manager, having completed 31 years with the company. Tony is well known to many of our customers as the person who advises them on their summer bedding requirements.

The oil-fired sectional boilers that Ralph and I had installed in the early 60s were giving us quite a lot of trouble and I spent many sleepless nights trying to sort them out. The alarm bells were now wired into our bungalow, so that at any time, night or day, I was alerted if the boilers failed. We also had a further alarm fitted so that if we had a power failure, I also knew about that.

I can't say that Hazel was always pleased when these alarms rang in the middle of the night, but she knew that they were vital for the well-being of the plants. I was therefore delighted when in 1976 the bank manager agreed to another loan, enabling us to install a brand-new boiler. In those early days, when cash was so short, we had always made do with the cheapest possible option for any of our plant and machinery, so it was pure joy to be able to install a boiler that was brand new and the very latest in design. One of the features that attracted me to this particular make was its ability to generate steam at the flick of a switch. I had experienced the process of steaming at Oaklands and thought that the level of sterilization it offered was greatly superior to the chemical method we were using which was time consuming and labour intensive.

The installation company recommended in their quotation that we order just one boiler, because they were so reliable. So we went ahead and took out the old sectional boilers, in their place a single Cradley Cornishman water/steam boiler was installed giving us hot water in winter and steam in summer.

Steam sheet method

The new Cradley boiler

It was a tragedy when this wonderful piece of equipment didn't live up to my expectations, being prone to all sorts of teething troubles. Once again I became accustomed to spending my nights in the boiler house, trying to keep the temperature in the glasshouses at an acceptable level. Searching for an answer to this problem, I phoned a heating engineer for advice. Although he could not sort out my various boiler problems, he did have a very good second-hand boiler, of adequate size, to offer me as a standby. The deal was done; the boiler was delivered and put in position. I learnt yet another very valuable lesson and since that day any piece of equipment that is absolutely essential to the smooth running of our business has a back up system in place.

Winter boiler problems became fewer. I was able to sleep more easily, but even so there were occasions when the heating did go wrong and I would drag myself up to the nursery. The boiler alarm would ring in the bungalow and at that time the only way to turn it off was by a switch situated in the boiler house. I shall never forget one occasion when the boiler alarm rang in the middle of the night. I didn't think it was going to be much trouble to get it started again, so I quickly pulled a coat over my pyjamas. Taking the dogs with me, I ran out of the back door and, realising that I had forgotten to pick up my keys, thought there'd be no problem with leaving the back door open. When I arrived at the boiler house, I turned off the bell and focussed my attention on sorting out the problem. It must have taken longer than I had envisaged and I never gave a thought to that 'open back door'. Twenty minutes later, I was alerted by frantic barking from the two very patient dogs and was somewhat surprised to find a PC Plodd standing in the

doorway. I don't know which of us was more startled. I was certainly frightened out of my life, but he boldly went on to explain that he felt he had to come and find me, to apologise for frightening my wife. Perhaps he was just checking out her story. He went on to explain his presence. He had heard a bell ringing and thinking it was a burglar alarm, went to investigate. On finding the back door open, he had gone inside calling 'Anybody there? This is the police!'

The children were fast asleep and Hazel, hearing a strange male voice coming nearer and nearer, but unable to decipher what it was saying, was terrified. Plucking up courage she had opened the bedroom door, clad only in her nightdress, but ready to defend her children. She almost passed out when she found a rather large policeman in her hallway. PC Plodd was very embarrassed and as you can imagine, when I got back to the house my wife was not a happy bunny.

... Plucking up courage she had opened the bedroom door

Very soon after that a switch was fitted that could turn off the alarm from the bungalow. The two boilers became very dependable and kept the nursery heated for the next twenty years. During that time the price of oil rocketed and we converted them to gas. This was a much cleaner heating medium and proved to be very reliable, but at the time we were very cautious and kept one of our old oil tanks full just in case we should need to convert back.

1977 was the Queen's Jubilee year and throughout the country there was a feeling of well being. The St. Albans Round Table were asked by the City Council to organise a carnival procession on August Bank Holiday Monday, starting at Bernard's Heath and passing down St. Peters Street and Holywell Hill, to arrive at Westminster Lodge by the lake below the Abbey. As I had been a member of Round Table and I very much wanted to celebrate this event, I agreed to participate. As usual, if I was going to do something, I was going to do it in a way that would create an impact – that was our style. So having agreed, I was left with the dilemma of what form our float should take. This was particularly important, as the organisers had now asked us to lead the procession.

We had participated in the show gardens at Springfields with planting beds of dahlias and we had made some good friends over the years. Every May they invited us to view the famous Spalding Flower Parade. We never managed to attend, but had always been interested in looking at the photos later in the year and been consistently amazed by the vastness and the intricacy of the floats. The idea came to us to ask Maurice Chapman, who at the time was chairman of

Springfields, for his expert help. He suggested that we paid a visit, because they still had a number of floats from the previous May's procession waiting to be dismantled in their yard. We were very welcome to use one of these, providing we could arrange the transport to St. Albans. I had a good friend in Mr. Jacques, who ran a local transport organisation and he promised to arrange this when required.

So Hazel and I set off for Spalding to view the floats and immediately spotted one that we thought ideal. It consisted of a coach, complete with six black horses, and had been the Lloyds Bank entry in the parade. Lloyds were our bankers, I assumed that as it was such a unique opportunity, they would want to sponsor us. How wrong I was. They expressed no interest whatsoever in our project, quite typical at that time, but what an opportunity they passed over. We were not to be deterred and decided to go it alone, but reduced the black horses to just two. Mr. Jacques suggested a low loader to bring the float back to St. Albans, but of course it was not as easy as that. We had to notify police stations along the route that we were bringing an extra wide load on a certain day. When it arrived, it had suffered slightly, having lost some of its straw foundation along the way.

The week before the parade, Hazel and the children patched the holes with straw. Any free time they had was taken up in making repairs. Then they set about covering the base with cupressus, to make a firm foundation ready for the dahlias to be attached. On Saturday morning, 5,000 dahlia blooms were cut and graded by colour. They were then put into flower buckets for a long drink. At 9 o'clock on the Sunday morning, an army of helpers arrived. This included any staff who could be spared from their duties, as well as a few friends who had volunteered to lend a hand. It was a mammoth task. We started by attaching with wire pins all the golden blooms on the top of the coach. We had built scaffolding around the float to help us climb to the top and had hoped to complete the coach using nothing but golden colours. But we very soon realized this was not going to be possible. A cutting team was sent out to strip the dahlia field of flowers and some of our preconceived ideas had to be modified. It was a glorious day, good for working in, but not good for the flowers. At lunchtime, more scaffolding was erected to go over the top of the coach, so that we could cover it in black polythene to shade the wilting dahlia blooms. As the day progressed, more members of staff arrived to lend a hand. By 6 o'clock in the evening we were more or less finished, although we still had the base to complete. Every flower that we could lay our hands on was used; we even raided the flower shop for the few chrysanthemums they had. When we just couldn't find any more flowers to add, we declared ourselves finished and stopped for a bite to eat.

There was an air of excitement when we stepped back to view the completed float and although we could have done with a few more flowers, we were satisfied with our achievement. It was a great day and there was a very good feeling of camaraderie, with staff giving their time freely, pleased to be involved in this once-in-a-lifetime event. The final task, before we all disbanded at 9 o'clock, was to push the float to the front of the shop, ready to be hitched to a tractor owned and driven by David Burrows, a farmer friend of mine and a fellow Round Tabler.

It had been decided that Julie and Sally, along with Mr. Jacques' children and other members of staff, would sit on the float on its way though the town. They would all be dressed up as flower girls, holding baskets of blooms that they could throw into the crowds.

All the floats assembled on the heath, ready to be judged before the grand procession and there was an air of excitement as people viewed them. This was the first time St. Albans had seen a floral float and it certainly made an impact. The appointed time came to set off and the flower girls took up their positions sitting all along the side. The bands started up and then near disaster struck. As the float bumped down the kerb, its tyres burst. It must have been the added

Taking a welcome break

Ready for the off

weight of the flower girls! Leaving only the children seated, everyone else decided that it was safer to walk. They soon found collecting tins to rattle for contributions to local charities. The memory of that fantastic day will stay with us all for the rest of our lives. Happy crowds lined the route and as we passed, spontaneous clapping broke out, bringing a real lump to our throats.

In 1980, when the nursery was celebrating its 25th anniversary, we decided to repeat the carnival adventure. This time we were asked by the organisers to carry the Carnival Queen, who had been chosen by the Review paper. Again, Springfields had a suitable float. It took the form of a stairway, leading up to a seat where the queen could sit. We called it our 'Stairway to Success'. This time we had better knowledge of how to construct the float, but the weather was not quite so good and the dahlias were less plentiful. We had to buy additional white chrysanthemum blooms to supplement our supply. Again the staff dressed up, but this time no chances were taken and only when the float had bumped down the kerb were the flower girls allowed to take their places along side the carnival queen. Although it was a great success, that magical Jubilee feeling was somehow lacking.

Aylett's GOLD MEDAL DAHLIAS
AUTUMN 1979 SPRING 1980 **JUBILEE EDITION 1955-1980**
GOLD MEDAL WINNERS RHS AUTUMN SHOW EVERY YEAR SINCE 1961

JUDITH

25th anniversary brochure cover

The team at St. Bernards Heath
Kathy Moxom and Eileen Samps
Sally Aylett
Julie Aylett
The float on its way
Stairway to Success

... the memory of that fantastic
day will stay with us all for the
rest of our lives.

149

We did however join the carnival for one last occasion. This time we were asked to sponsor the Household Cavalry's appearance. To advertise our involvement, we decorated one of our vans with dahlias, which Brian then drove at the head of the procession through the streets of St Albans. We decided that the only way to participate in this annual event, is to do it in style. Somehow that happy combination of so many factors that made that first float possible have never arisen again.

Household Cavalry   Decorated van

chapter 11

... flower shows filled a large part of our lives

Flower shows filled a large part of our lives during the summer months. I have already mentioned how we had become involved with this form of retailing. As we attended the various flower shows, we became very friendly with our fellow exhibitors. Some of them were particularly kind and supportive to us. I well remember two Welshmen, Ernie and John, who were in charge of the Wyevale Rose exhibits. These two, along with Ernie's family, were always keen to help us and to advise on which shows they thought we should exhibit at. Ernie's kindness extended to giving us lodgings with his family when we went to the Royal Welsh Show. Before our return home, they gave us a tour of the Wyevale Nurseries at King's Acre who grew a vast selection of nursery stock.

Another character who was well loved by everyone in the flower show world was Charlie, who was Kelways Nurseries show expert, it was through him that we supplied a large quantity of dahlia plants to this well known nursery in Somerset. One year, Hazel drove the Land Rover and trailer loaded to the gunnels to Somerset, offloaded the plants, made her return journey and arrived back late in the evening. Yes, we certainly went to extreme lengths to sell our dahlias.

Another venture that we became involved in was with another famous nursery, Blackmore & Langdon. They brought together eleven other specialists to produce a 'Best of British Horticulture catalogue'. This was distributed to all the participants' mailing lists, offering a wide range of plants. It never really succeeded and was only published for the one year, but it was a great idea nonetheless. It was truly gratifying to hear these two nurseries mentioned in the BBC's coverage of the Chelsea Flower Show last year. Both are independently owned and Kelways still specializes in peonies, while Blackmore & Langdon is still run by the original family and received a Gold Medal at Chelsea in 2004. There was always plenty of time to gossip with each other at flower shows, as we waited for the crowds to arrive. We learnt to recognise those characters who would inflate the numbers of orders that they'd taken, never admitting that their order books weren't full. We would deduct at least fifty percent from their total to estimate the real figure. But you could always listen and learn which shows were successful and worth giving a try.

Gradually the number of shows we were attending grew to about 24 and when you consider that these took place in a matter of 14 weeks, you get an idea of the workload. At the height of our showing days we travelled as far afield as Builth Wells for the Royal Welsh Show and up to Derbyshire for the biggest one day show, held at Bakewell and to an unforgettably awful show at Blackpool, where we managed to take all of a dozen orders.

Flower shows were certainly a very good way of getting to know the whole of Britain. Hazel or I would drive the Land Rover and trailer containing the show equipment to the showground and the flowers would arrive by hired lorry. Hazel didn't like backing the two-wheel trailer, but apart from that she was quite happy driving. During the latter part of the 70s, we finally outgrew the Land Rover and trailer and concluded that the business had grown sufficiently to support the purchase of a lorry that was capable of providing all our transport needs.

Brian already had his licence to drive Group 3 vehicles, but I didn't, so I signed up for a few lessons with the British School of Motoring and attended the heavy goods vehicle testing station. To my relief I passed the test and in the process of doing so learnt an immense amount of very valuable information on how to handle a lorry efficiently.

These shows were self-funding because the prize money, added to the sale of the flowers, would usually cover our expenses. They were a very good way of spreading the name of Aylett's Dahlias around the country. We grew a selection of varieties as cut flowers under glass so that we could try the early shows. We attended the Royal Agricultural Show at Stoneleigh, as we discovered that the farmer's wives were the gardeners and didn't mind spending their husband's cash while the men folk were busy drooling over the latest design in tractors.

Our first lorry

Although these early shows were good for the order book, they were not really financially viable and we only participated for a few years. At the other end of the dahlia season, we found another show that we thought the farmers' wives would attend, namely the Dairy Show at Olympia, in London. The horticultural section was upstairs and we had to cart our equipment, plus buckets of dahlias in a lift. We were exhausted before we even started arranging the flowers.

It was at this show that the Queen Mother, on her meandering through the stands, stopped and showed an interest in a soft salmon red dahlia named 'Muriel Gladwell'. I had often dreamed of holding a royal warrant as a supplier of dahlia plants, having always admired the royal crest on Stevens the rose growers, entrance in the Lea Valley. I thought this was my chance and on my return to the nursery a very large bouquet of this dahlia was prepared, which I duly delivered to Clarence House in the hope of an order for some plants. Although I was to be disappointed, I did receive a very nice letter from Her Majesty's Lady-in-Waiting thanking me for my trouble. It was a letter that I treasure to this day.

CLARENCE HOUSE
S.W.1

31st October, 1966

Dear Mr. Aylett,

I am commanded by Queen Elizabeth The Queen Mother to thank you for so kindly sending a bunch of the dahlia "Muriel Gladwell", which Her Majesty had admired on your stand at the Diary Show last week.

The flowers are now in Queen Elizabeth's sitting room, and are, I know, giving Her Majesty great pleasure, they are such a delightful colour and shape.

Yours sincerely,

Olivia Mulholland

Lady-in-Waiting

R.S. Aylett, Esq.,
Aylett Nurseries Limited.

Letter from the Queen Mother

chapter 11

We had many adventures during our showing years and I think we gained a reputation for being a pretty scatty team. Although it was a very exhausting lifestyle, it was great fun and we laughed a lot and never louder than when we were at Worcester. We were very late staging, which was normal for us and the hotel had given us a key to the front door. We duly arrived at 2am. All was very quiet and after letting ourselves in, we crept upstairs said goodnight to William. Moments later there was a great commotion because William, thinking he had turned on the light in his room, had actually hit the fire alarm button. All hell broke loose, and we must have woken the whole hotel. Needless to say, the proprietors didn't think it was at all funny and we never went back to that hotel again.

Another show that I particularly remember because of its trauma was Southport. This show had the reputation of being the Chelsea of the north. It was a 3-day show, which meant that we had to pick the dahlias on the morning of our departure, so that they would remain fresh. Hazel should have been on the road, driving the Land Rover in the morning, but as luck would have it we had problems with the vehicle. Eventually, after a further breakdown on the M1, we arrived at Southport at two o'clock in the morning. We were booked into bed-and-breakfast accommodation in Ainsdale, a district of Southport. After a very short night, we rolled out of bed and staggered downstairs feeling a little under the weather, to say the least. The kindly landlady greeted us with an enormous plate of kidneys, bacon, tomatoes, eggs, fried bread and, to crown it all, black pudding. The girls went deathly white and not wishing to upset our very caring host, a quick search was made for a receptacle to accommodate the said breakfast. A paper bag was found and the offending repast was disposed of; we bade our farewells to our landlady, assuring her of our delight in her delicious food, which was soon deposited in a waste bin.

Show officials tended to be a breed of their own. Some were extremely pleasant and helpful and others were full of their own importance and determined to see that the rules were followed at all cost. Southport had more of the latter than most. It became one of my least favourite shows, as we seemed pre-destined to upset the organisers. It had, according to Hazel, two redeeming factors. One was that she had tremendous fun driving the Land Rover along the sands to the showground and the other was that William's uncle lived nearby and treated them both to a superb dinner at the best hotel in town.

At the end of the show, the dahlias were a sorry sight and it was decided that they were not fit for selling. Before the dismantling of stands could begin, a bell was rung and, as usual, a large crowd had gathered around. On finding that we were not willing to sell the dead flowers, poor Hazel and William were mobbed and almost trampled to death in a mad rush of people determined to help themselves. They never again refused to sell dead dahlias, although for the life of them they couldn't understand why people were willing to buy dead, rotting blooms that were only fit for the compost heap.

I have already mentioned that our favourite show was Shrewsbury, We had graduated from sleeping in a tent and hired a caravan for our accommodation. This would be towed into position alongside the River Severn. It was almost like a small village of exhibitors and the society looked after us well, providing standpipes at intervals alongside the riverbank and superb toilet facilities.

One year, one of the star attractions in the show arena was a display of the Queen's Own Household Cavalry. They were camped alongside the riverbank, complete with stabling for their horses and early in the morning they would ride past us, which was a truly magnificent sight. One morning I was out on the riverbank cleaning my teeth and as they went by, the captain, on his beautifully groomed horse, called out 'Lost your bathroom, Sir?'

... Lost your bathroom, Sir?

As the children grew older they were allowed to accompany us. We then hired two caravans. Later on, we became more adventurous and actually purchased a second-hand one of our own. Hazel towed this behind the family car and would leave a day earlier than the team in order to prepare the bowls with the wet peat.

One year, we decided to enter three extra classes in the smaller marquees, in order to try to boost our prize money. This was in addition to our 40ft. stand in the Quarry Marquee that housed the crème de la crème. One of these additional entries was a floristry exhibit, which had to depict a theme. Hazel decided to choose the Changing of the Guards as her subject. Unfortunately, the night before staging, she was struck down with a tummy bug. We all thought would have to abandon the entry, but she made a speedy recovery and went on to win first prize. In fact, we won top awards for all our entries and went home very pleased with ourselves, but we never repeated the exercise. It was just taking on too much.

We won many trophies at Shrewsbury over the years. The competition was always fierce, so the winning was all the sweeter. There were also highly desirable pieces of china and glass awarded,

Eileen, Sally, Hazel and Julie with the gold cup at Shrewsbury

Show preparations with peat bowls in foreground

for the lucky winners to keep. I remember Hazel phoning with the news that we had won the very prestigious Everest Gold Trophy. With it came a rather ornate Royal Worcester china coffee pot, which Hazel didn't much like. As she walked back to the stand, very nonchalantly swinging the coffee pot on her little finger, she was approached by a bystander who offered her £50 for the said piece, at which point she quickly changed her mind about it and took a little more care.

It was also at Shrewsbury that we introduced a unique dahlia that Mr. Cooper had named after my daughter Julie. We arranged over two hundred stems on their own and they certainly made quite an impact. Today, this variety is the only 'Jescot' dahlia that we still grow. In 1975 'Jescot Julie' was awarded a Highly Commended Certificate jointly by the R.H.S.and the National Dahlia Society.

Dahlia 'Jescot Julie'

Various trophies, gold medals and pieces of china and glass won over the years, note the coffee pot from Shrewsbury in the centre.

EX

An Island exhibit in the R.H.S. new hall at Vincent Square

In London, we attended the fortnightly flower shows of the Royal Horticultural Society held at their halls in Vincent Square, Westminster. These halls were also where the National Dahlia Society held their annual show. This event attracted dahlia enthusiasts from all over the country; they would arrive by the vanloads

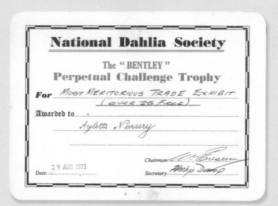

in the evening, with their carefully cosseted flowers often wrapped in cotton wool. We would usually be putting the finishing touches to our own exhibit at this time, clearing up and ready to drive back to St. Albans. They would happily work all night, arranging their dahlias most meticulously, whilst keeping a close eye on the competition.

We were viewed with a fair amount of suspicion as we grew quite a few dahlias to please the flower-arranging ladies. We had also broken with the traditional, multicoloured, wallpaper style of the trade stands. We arranged our dahlias in wet peat too, and in doing so created a fair amount of mess.

When we won the top award, the Bentley Trophy, for the first time in 1970, we realized that we had finally been accepted. Sadly, as the years went by and dahlia growers became fewer and fewer, we were left as the sole trade stand. The reason for this decline, in my opinion, was brought about by the show being

Bentley Trophy certificate which we were awarded for 5 consecutive years

downsized from a 2-day show to a 1-day, which then did then not warrant the tremendous expense of staging an exhibit in the middle of London. Secondly, parking became an absolute nightmare. And thirdly, there was a definite decline in the business to be had. Dahlia enthusiasts usually swap their plants and only actually buy the new varieties.

We had been winning gold medals at the R.H.S. shows since 1961. At about this time, we were invited by the Army & Navy Stores in Victoria to set up an exhibit of dahlias on their premises. This was to be staged to coincide with the R.H.S.'s Great Autumn Show. The idea was that people arriving at Victoria Station would have to walk past the window, en route to the show. Seeing the dahlias, they would be drawn into the store and become potential customers.

Display at the Army & Navy Stores, Victoria, London

As we were exhibiting at the show itself, this dual commitment meant an early start to the day. We would leave St. Albans at 5 o'clock in the morning, picking up Hazel's Mother on the way. She lived in Hendon at the time. The store was unlocked at 6 o'clock and we started what for us was a very long day. Both shows were highly successful and the store was so pleased with the increased sales that we began exhibiting at many of the Army & Navy's branches in the south of England. In Guildford, I remember arranging dahlias adjacent to the china department, which was particularly nerve-wracking. I had this nightmare vision of shelves of china crashing to the floor. Luckily for us this never happened, but we did have to polish up our staging techniques. Wet peat in a flower show is acceptable, but in an upmarket department store it's a definite no-no.

The next store to approach us was a well-established firm, Carter Page. They had a very old fashioned shop situated at London Wall, in the City of London. The store catered for the City's gentlemen who, during the lunch hour, would dash in to make their purchases. It was a completely self-contained community and one that time seemed to have passed by completely. It seemed really strange that we were in the very centre of our capital, but there were none of the stresses and strains of busy, everyday shop life. We enjoyed our time there. Eventually, Carter Page was taken over by Bypass Nurseries, who supplied their own dahlias and obviously dispensed with our services.

Display at the R.H.S. Great Autumn Show

The Gardening Centre at Syon Park was sponsored by I.C.I. and in 1968 opened show gardens. Here different specialists supplied the plants that made up the displays, which the public paid to view. Percy Thrower was involved in the project and there were many different gardens within the grounds. One was called the Electric Garden and there, at the flick of a switch, all the chores of gardening could be completed automatically. The grounds also had a newly constructed conference hall and it was to be here that we were invited to stage an exhibit of dahlias. A so-called service sheet was handed out during the show; it was really a leaflet all about the centre. On the front cover was a picture of Hazel, holding the cup that we had recently been awarded for the best exhibit at the Birmingham Flower Show. The whole project never really took off, as it failed to gain popularity with the public and I.C.I. had had to rethink their idea.

During the years that we exhibited at the R.H.S. shows, both The Daily Telegraph and The Times were invariably very kind to us. Normally on a Wednesday morning, after an exhibition at the Horticultural Halls, they wrote up the details of our exhibit and suggested that readers should go along to see the show. I know this sounds a little arrogant, but by the 70s we would have been a little upset if we'd been awarded anything less than a gold medal at the R.H.S. shows. But there was a special award that we'd now set our hearts on and winning it became a special challenge to us. As each exhibit was completed, Hazel and I would look at each other and there was always an unspoken question between us. Nobody else would know what we were thinking. We were our own most critical judges, knowing when the dahlias were of a high standard and knowing when our exhibiting skills had done them justice. The award that we always dreamed of winning was the Williams Memorial Medal. It was given for the best exhibit of plants and/or cut flowers of a single genus, staged at one of the society's shows during the course of the year.

At the Great Autumn Show in 1977 our dream came true and I had the honour of being presented with the Williams Medal by Lord Aberconway, at the annual general meeting the following February. Dahlias were not his Lordship's favourite flower and in his address he made note of the fact that Dr. Dahl would not of been pleased with the mis-pronunciation of the genus. With this statement he made quite sure that we knew our position within his horticultural world, but his remarks did nothing to diminish our great sense of achievement. We were to win this award on two further occasions, before deciding in 1999 that exhibiting at flower shows had become too expensive to continue. It was also becoming a strain finding the right personnel for this very specialist art.

Williams memorial presentation and below the winning display

As a result of all our exhibiting at the Horticultural Halls, I was invited to join the Joint Dahlias Committee representing the Royal Horticultural Society. I served on this committee for 30 years, holding the position of joint chairman for 12 years before retiring at the age of 70. One of the duties of this Committee was to look after the dahlia trials at the R.H.S. Garden at Wisley. This involves the members of the committee meeting six or seven times a year during the flowering season. They also meet in five different parts of the country, to select new varieties that various breeders put forward, in the hope that they will be chosen for the following year's trial.

In the early days, the journey to Wisley was long and tedious, via Staines and Chertsey and taking a good two and a half hours. The opening of the M25 changed all that and 40 minutes on a good day is all it takes today. During the school holidays, I would be accompanied by Hazel, the children and, in due course, the grandchildren. They all loved looking around the gardens, but the highlight was definitely the promise of Honey Crisp, a speciality of the people who ran the excellent restaurant in those days. Sadly, this is no longer available, much to the regret of our grandchildren. Thomas, my eldest grandchild, was very disappointed when I retired from the committee at the end of 2003. The thought of not being taken to see his beloved model vegetable garden was almost too much for him to bear. I have made him the promise of an annual visit to check that all is well.

A plate presented to Roger Aylett on his retirement from the Joint Dahlias Committee

chapter 11

Wisley has changed in many respects. In my early years there were some 200 varieties on trial whereas recently, with the demand for trials of other plants, there was only space for about 145 varieties. The coveted Award of Garden Merit – the A.G.M., for short – has superseded the old awards. This is only given to varieties that are grown for more than one year in the trials, show excellence in performance, are an improvement on a existing variety, or completely unique in form. During my time on this committee I made many friends and my knowledge and understanding grew about many aspects of the cultivars being grown in the trials.

Roger Aylett and his grandson Thomas at Wisley 2003

In a personal capacity, my other involvement with the R.H.S. was serving on the shows committee. In the early days, this involved attending meetings to discuss various issues brought to the attention of the committee by the staff of the R.H.S. The committee was then reformed and one of its new duties included the selection and assessing of the sundries stands at the Chelsea Flower Show.

This wasn't quite how I had envisaged making my comeback to Chelsea, but it's a job that has given me a great deal of interest and pleasure. In June 1999 I had the honour of being presented by the R.H.S. with the Gold Veitch Memorial Medal. This is conferred on persons who have helped in the advancement and improvement of the science and practice of horticulture, and represents the pinnacle of my career. I felt very humbled to be amongst a favoured few.

The same year, the National Dahlia Society conferred on me a Silver Medal for my services to the dahlia.

Presentation of the Gold Veitch Memorial Medal by Sir Simon Hornby V.M.H.

...we were told to continue with our traditional practice

In about 1981, a friend of mine who was in the computer industry, told me of a grant of £2000 that was available on his recommendation to companies who had not advanced into computerisation. This allowed his company to do a feasibility study, conducted on site, which would determine whether we should invest in a computer or carry on with our existing methods. At the end of this study we were told to continue with our traditional practice, as they could not improve on the way we had been operating.

Julie, while studying for her A-levels, had asked if she could join me at a trade show I was going to the next day. This exhibition was British Growers Look Ahead, held at Harrogate. I had always found my annual visits to B.G.L.A. very worthwhile, with plenty of new and innovative ideas on display for the horticultural industry. We found lots to interest us and on the drive home Julie asked me, 'What was the most interesting thing you have seen today?' I replied that it had to be a little computer that printed labels for certain plants. When we arrived home, we told Hazel all about it and it was agreed we would ask the representative to call and set up a demonstration.

The company selling this labelling system, Lebet Ltd. turned out to belong to none other than, the late Leith Hayes, who ran a marvellous garden centre at Ambleside, in the Lake District. Leith was a real character and had the foresight and intelligence to see that the garden centre industry was about to take off and that the old practice of hand writing labels was going to be a stumbling block. Accompanied by his wife Betty, Leith duly arrived to set up the demonstration, which took place during the morning. It was fascinating we felt the computer would be of help to us, but we hadn't the confidence to go ahead without talking it over with our members of staff, after all they were the people who would operate the system. After relaying this to Leith he replied, 'I would rather demonstrate it to them this afternoon.' At the end of the afternoon, or I should say early evening, Leith said he must make his way home, as he had an appointment early the next morning. We said he couldn't possibly drive home the same night, after such a long day and he was persuaded to stay the night with us. We still hadn't made up our minds to purchase the system, but Leith very forcefully told us he certainly wasn't taking the machine away and we were to keep it on a trial basis. He left early the next morning, minus the computer, to drive back to Ambleside. Needless to say, we bought the system. Ironically, it had been less than two years since the feasibility study we had undertaken stated we were not ready for 'computerization'.

The system was run on a Superbrain computer with a top capacity of 700k. When you consider that just one of our digital photos today is 1762k, it shows just how far the computer industry has moved. The data was stored on 5in floppy disks, which were accessed from drive B, while the operating system was stored on drive A. The two drives had to work in harmony, but, for all kinds of reasons, they frequently didn't. We must have driven Leith mad, as we were constantly on the telephone with problems. In time all but Hazel had given up on the Superbrain computer and the system resided in our spare room. Hazel, with a lot of help from Leith, finally cracked it and became proficient. Hand-written labels became a thing of the past.

At that time, Hazel was processing the weekly wages using a Kalamazoo manual system. It was long winded and took a whole Thursday to complete. In April, at the end of the tax year, when all the p60s and returns for P.A.Y.E. and N.I. had to be completed, a whole week had to be allocated. So when Leith mentioned to her that the Superbrain computer was capable of running other programs and Hazel found out that one of them was wages, she was enthusiastic about finding out more information on the subject.

I phoned Superbrain, who recommended a certain Mr. Peter Daly from Northwood. He had written a package of accounting programs that could be run on their machines. We phoned for an appointment and went to his office in Northwood for a demonstration of his wages package. It was all Greek to me and Hazel admitted on the way home that she hadn't really understood a word, but was willing to give it a go. Eventually, Week 1 of the wages year arrived and it was decided that Hazel plus the Superbrain would go to Northwood to work under Peter's supervision until she felt confident. From day one they seemed to be able to work with each other and this happy working partnership has continued to the present day. After we had conquered the labelling and the wages system, Peter suggested that we should look at the 15,000 names and addresses that made up our mailing list and were typed out twice yearly. Today our mailing list numbers over 35,000 names and addresses all processed by the computer.

We had upgraded our Superbrain by then, investing in a networked system supplied by a local company based at Hendon that Peter had recommended. By now we had achieved some understanding of computers and the benefits of using them, so when Peter suggested that our stock of plants on the labelling system could easily be turned into a stock control one using the existing data, we agreed.

Some years later we turned our attention to what we call the 'sharp end' of the business and again Peter acted as our consultant, accompanying us to west London to look at some new point-of-sale computerized cash registers. The representative was very proud of his machines and told us that they were capable of storing 25,000 products. As we had some 40,000 items on our system, we thanked him very much and made our way home. Peter was very quiet on the way back and suddenly said to us, ' I think we'll build our own system.'

We had been noticing that hardware goods were arriving from the wholesaler with strange looking lines and numbers on the packaging. We learnt that these were called barcodes and were a method of identifying items and this was to be the way in which retailing would be taking advantage of electronic point-of-sale capabilities. At that time, about fifty percent of hardware was barcoded, but plants were a different subject and no one was even thinking about it. Even Leith thought it was a mad idea and far too adventurous for the horticultural industry. Nevertheless, we decided that if we were going to build our own tills, we would start barcoding our plants, so that when the time came, we would have the capability of making full use of the technology available. This in itself was no easy task, as inks and

'the till girl just points what looks like a hair drier come torch at the barcode and like lightening the product and its price lights up on the till'

All eyes on Aylett's EPOS

THE INS AND OUTS of installing and operating an electronic point of sale till system were the topic of an HTA afternoon hosted last month by Roger Aylett at his nursery/garden centre complex at London Colney. Peter Daly, who has been responsible for the design and installation of the system now on-line at Ayletts, explained the thinking behind it. Visitors then had a chance to see at close quarters members of Aylett's staff putting the demonstration equipment through its paces and see it in operation in the garden centre.

NURSERYMAN & GARDENCENTRE 9 JULY 1987

By Peter Seabrook

Nurseryman & GardenCentre

Next day it was Ayletts and the HTA afternoon concentrating on Micromas Tillmaster computers. This system at London Colney really is most impressive and one you must see. The till girl just points what looks like a hair drier come torch at the bar code and like lightening the product and its price lights up on the till. A wipe of the credit card through the till slot, a whirr of the electronics, a signature and the customer is on his way with a very detailed receipt.

It's one drawback — cheques slow down the proceedings!

chapter 12

176

labels had to be researched in detail as the barcode readers had to be able to decipher the code, even after it had been exposed to the elements.

During her holidays from university, Julie tried out the prototype for Peter. It consisted of a V.D.U. with a swipe reader stuck to the side, a Symbol laser barcode reader, a printer and a cash drawer. Eventually, in the spring of 1987, the E.P.O.S. system was ready to try out on our valued customers. It was a great success. The queues, which had been horrendous, disappeared, while the staff found it easier to process a trolley-load of items. The customers knew exactly what they had been charged for and stock became much easier to order, as we knew exactly how many plants we were selling, or packets of slug pellets we had on the shelves.

The Horticultural Trades Association heard of our new achievement and asked if they could have an open day so that the system could be shown to the trade. On 23rd June, we set up a trial unit in the potting shed and demonstrated the system to interested garden centre operators. Members of the trade and press showed a great deal of interest, but I had the distinct impression that they thought we were quite mad in making such an investment. But we were happy and these home made tills served us well for many years and eventually the rest of the trade caught up with us.

One of the disadvantages of going ahead on our own was that we had our own unique system, purpose-built just for us, so when the rest of the trade finally got round to realizing the advantages of stock control and electronic point-of-sale,

Our own designed tills capable of reading barcodes

we were left isolated. We were unable to take advantage of any enhancements the bank clearing houses made, unless we implemented them ourselves with new programs and rewrites. Realising that this was not a very sensible position to be in, we enlisted the help of Peter Daly in writing a program that could export our stock codes into a format so enabling us to purchase electronic tills from a source that supply many garden centre outlets. This process is currently used in our till system today. Barcoding has come to be regarded as an industry standard in the modern garden centre trade. However, at the time of writing this book, we are using our original, old reliable stock system – we're very reluctant to give it up, although the time is ticking away and I know it will eventually have to be updated.

It was during another of my visits to the British Growers Look Ahead exhibition that I first encountered a computer system which I thought would definitely improve efficiency in terms of the running of the nursery. It was at this annual venue that I had first seen so many good ideas that we subsequently purchased and installed. On this particular occasion, I was very impressed with a computer that could open and close glasshouse vents automatically, as well as operating the thermal screens. It was able to set various zones in different greenhouses to different temperatures, by the opening and closing of the heating valves as the set points dictated. It was also able to operate the irrigation system. This again could control designated zones, so that the plants were watered throughout the night. All these tasks were carried out manually at that time, and the very thought of not having to spend my summer evenings going backwards and forwards to the nursery, turning one row of irrigation off before the next row could be turned on, filled me with pure joyful anticipation.

The firm that had been exhibiting the equipment at B.G.L.A. recommended a specialist, based in Harlow, who was experienced in installing such systems. Bruce & Sons is a family concern, with a vast knowledge of glasshouse environmental control. Mr. Bruce Senior was the first member of the family I had the pleasure to meet and it was he who explained what was initially required to prepare the site before we could even think of installing a computer. This involved making alterations to our heating pipes, installing various pumps and mixing valves, thus creating the zones that could be identified by the computer. These ten individual zones can all be set to have different temperatures and the computer then sends the appropriate instruction to the boilers.

The other massive task was the installation of the cabling that would carry the information back to the office, where the nerve centre was to be situated.

chapter 12

Julie and Roger Aylett with the environmental computer

The preparatory work was duly carried out and the day arrived for the computer to be installed and the system tested. We now had the vents opening and shutting as the climate dictated. Wind speed is measured by a sophisticated weather station. If the wind is too strong, the vents are automatically lowered.

A terminal connected to the main computer was installed in the bungalow. I could now oversee the boiler operation without leaving the comfort of my home. Unless, of course, there is a mechanical problem, because a computer that can go and physically fix all boiler house malfunctions has yet to be invented.

... the boiler operation without leaving the comfort of my own home

... by the 80's the nursery had become a success

By the 80s, the nursery had become a success, although not as I had planned it. In reality, it had evolved as circumstances had dictated and had become quite diverse. There were now four separate businesses under one company name. We had a very busy flower shop, with an excellent reputation employing, at least six people. We had a highly successful Gold Medal dahlia trade, a wholesale cut flower and nursery business and an ever-expanding garden centre. This was enough to keep me fully employed and I worked a seven-day week. I didn't mind, because working at something you love is not a chore, it's a way of life.

Hazel was also working full time and would swap from job to job as the need arose. We always managed a two-week family holiday in late July. In the early days, Chips would invariably accompany us to our chosen holiday home. He was not about to be left behind and would lie behind the car until his allocated space was arranged. Hazel's parents would take over the day-to-day running of the nursery, with Brian as their right hand man, keeping in touch with us daily to discuss any problems. Later on, as Chips became too old to make the journey, they had the additional task of dog sitting. When the realization struck the family that Chips was not immortal and was coming to the end of his life, a new Labrador puppy named Jemma arrived. Chips gained a new lease of life and had another two years with us before the sad day arrived when we had to say goodbye.

It was during one holiday in Devon that we stayed in a cottage attached to a farm. The farmer's wife had a quantity of chicks and ducklings that she was raising, with the sole idea of popping them into the freezer for the family's dinner table.

We had recently bought ourselves a very large chest freezer that needed filling, so I paid particular attention to this production line. On our return home I made a trip to a farm in Codicote, where I knew they sold day-old chicks. The only ducks they had were Khaki Campbells and Cobbs were the only chickens they could offer. In my ignorance, I didn't think that mattered too much, and I arrived home with twenty-four of each. Hazel always says I only think in double figures. Unfortunately, neither breed was suitable for the table, as proved when the first duck matured. They did, however, lay an enormous quantity of really excellent eggs that we were able to sell in the shop. The ducks, having outgrown the garage and garden, graduated to the lawn on the front of the nursery. From here they would sneak into the dahlia field, to nest under the plants in the hope that they might just be overlooked and so hatch a clutch of ducklings. Unfortunately for them, they were soon discovered as we went about our daily task of picking the dahlia blooms.

I well remember one Sunday morning, when I was woken by the front door bell. On the doorstep I found a neighbour, who thought I ought to know that our ducks had crossed into the central reservation of the road, where they were happily feeding. Fearing an accident, I quickly dressed and ushered them to safety. We soon realized that having had the taste of freedom, the ducks wanted more, so in an effort to curtail their wanderings we took them to the very far end of the nursery where there is a pond. On finding the pond they all quacked with delight and took to the water. We made our way back, very pleased with this solution, but in less than five minutes the ducks had found their way back to the central reservation once more. Enough was enough. We found them a new home in Sandridge, where after a brief episode in which they escaped into the churchyard, they lived happily for many years.

The chickens had been ordered out of Hazel's garden, when she found them roosting on the front porch. They took up residence in the 'guinea pig' glasshouse for the winter, but unfortunately they came to a sorry end when a fox gained entry and carted them off. Ever since that day I nurtured a great dislike for foxes, as this particular one left a trail of chickens bodies across the field, in what I assumed was just wanton slaughter. It wasn't until I was watching Bill Oddie on television recently that I discovered that foxes kill and then bury their spoils for later consumption. I now realize that the fox had probably been disturbed before completing this task. However, I had decided that this brush with food production was enough for me and decided to stick to growing plants.

The shop, which had originally been established as an outlet for the flower crops that we grew in the early years, was now flourishing. It needed a greater selection of cut flowers than we could produce ourselves, so twice a week I would make the trip to Covent Garden to purchase additional supplies. Initially, Hazel would accompany me, but as the children arrived this became impractical, as baby sitters were hard to find at 4 o'clock in the morning. The old market at Covent Garden was a world of its own. I, of course, had knowledge of its workings from my point of view as a supplier of produce. I now experienced the other side, as a buyer. Our vehicle fleet had expanded to include a van and it was in this that I would arrive at market.

Different traders had their regular customers, who, on arrival, would often leave their vehicles in the middle of the road, with the keys in the ignition, to be parked and shunted around by their own particular self-styled personal assistance, mine was named Harry. Harry looked after me in Covent Garden and then at Nine Elms when the market moved there some 30 years ago, Harry retired before I gave up the market. The system worked admirably and the police were only too glad to have the traffic managed in this way. By 8 o'clock, the streets would return to normal, as the commuters started to arrive to occupy the office blocks, completely unaware of the dual life of the area.

Sadly, the day came when the old market could no longer cope with modern day trading and the whole operation was relocated to a purpose-built site, across the river at Nine Elms. This new market was sterile, with no real character of its own and the unions moved in and rules came into force about who could do what. We were allowed to move our own produce in, as long as it was before midnight. After this time it was a brave man who defied the union rules and didn't engage the services of a porter. The children's great treat was to be allowed to accompany me to market in their school holidays. They took particular note of the way I would negotiate the price with the salesman, a lesson they have both put to good use in their lives. Julie and Sally soon became very adept at loading the van and became friends with the porters, who were all very fascinated to see these two young folk coming to market at that time in the morning. We would arrive home in time to open the shop, often they would almost be completely engulfed by the boxes of flowers that had to be crammed in around them.

Having grown up on the nursery, the children were very involved in the development of the business and were glad to lend a hand with any task, but we never expected them to follow in our footsteps. It was a hard life, not for us bank holidays and long leisure hours – we even worked at weekends. But it was the life we had chosen and we certainly didn't see why our daughters should not do their own thing. This was made quite clear to them and in fact Hazel made quite a point of telling them that they should find different careers.

When Julie was studying for her O-Levels, she woke up one morning with the news from Hazel 'that we now had a women as Prime Minister and there was now no career that was not open to her.' Julie quickly retorted: 'I just want to come into the business and you won't let me!' From that moment on, we never tried to dissuade Julie from joining us. We were, however, determined that she should have the chance to go to university, so with that goal in mind, she studied the appropriate subjects to gain entrance to a degree course. Julie decided that Bath was to be her first university choice, the college at Wye being her second. However, she was to be disappointed when she did not get the grades that Bath had stipulated and was offered a place at Wye. The offer sat on her bedroom table for many days, as she tried to come to terms with what she considered to be a failure, but in the end she accepted and that marked the beginning of a wonderful three years for her. We enjoyed this part of her life almost as much as she did herself and found her new friends to be a very interesting group of young people. We began to notice that whenever she was recounting her latest escapades to us, the name of a certain young man would crop up more than others.

This was Adam Wigglesworth and a few years later we welcomed him into the family as our first son-in-law. During Julie's time at Wye, she furthered her experience by becoming Vice Chairman of the Students Union. She found this very interesting and during one Easter vacation went to the students' annual conference.

Julie graduated in June 1986. She now had letters after her name: BSc Hort. (Hons) and needless to say, she had two very proud parents present on her graduation day. Julie's college friends all met again for her 21st birthday some two weeks later, held in the same venue as my own 21st: the Water End Barn in St. Albans. To her delight, they replicated the cake that she had always admired in photos of my own cake. The evening was a great success and a fitting end to three happy years. During Julie's last year at Wye we had discussed what her plans were for the future. It was always the same answer: she wanted to 'come back and work at the nursery.'

However, she was keen to see a little of the world before she knuckled down, and had quite an adventure on a trip to India with a girlfriend from Wye. After this, it was time to start work and it was decided that she should widen her practical horticulture experience by joining a well-known houseplant nursery in the Lea Valley. This was not a success, the only experience she found of any value, lay in learning how to fend off the advances of amorous Italian fellow workers. Enough was enough and she decided that the time had come when she should follow her heart and return to the family business. Her desk was placed next to mine and working closely together she soon proved more than capable.

She was happy to become a director of the company in August 1987. Hazel and I felt that if Julie was to work with us, it was not right that she should also live with us. Children need independence from their parents and Julie, having lived away from home, had already flown the nest. We suggested that she should invest in a small cottage, not too far away in Wheathamstead and it was from here that she travelled to the nursery daily.

One Saturday morning in May 1989, when I was busy with a stream of customers, I suddenly found a very agitated Adam standing at my side in the shop. He asked if I could spare a moment in private with him, so we went up to my office where the dear old-fashioned boy asked for my daughter's hand in marriage. He tells me that I replied 'And about time, too!' In February of the following year we celebrated their marriage in St. Marys Church at South Mimms.

... when I was busy with a stream of customers

Our second daughter, Sally, made a sideways move from the world of horticulture. She decided after her A-Levels that agriculture and the wide-open spaces were for her. The college of her choice was Seale-Hayne at Newton Abbot, in Devon and one of the requirements was a year's practical experience. So she enrolled on a youth

training scheme at Oaklands, which meant that she could remain at home. Her least favourite duty was in the piggery. This was in the time of near-factory farming; the sows were kept in a very confined space and the piglets never saw the light of day. After a day of loading pigs into a lorry to go for slaughter, I remember asking her if she found it too distressing. Her answer was 'No, as the pigs didn't have much of a life anyway.' It was soon after that she became vegetarian and meat was off the menu.

One day she was very late home. We were relieved to hear her key in the door, but one look at her drained white face told us that something was wrong. Evidently she had been instructed to drive a tractor to the college's farm at Bayfordbury and this involved driving along the busy A414. This was the first time she had ever driven a tractor, let alone on a busy road, but in true Sally style she

Sally Aylett at the computer

chapter 13

managed it. I do have to admit that this left me thinking that the training on the Y.T.S. course left a lot to be desired. However, this did not deter her from farming and off she went to college in Devon. After gaining her Higher National Diploma in Agriculture, she returned to St. Albans for a short time before she too went off to India, where she met up for a day at the Taj Mahal with Julie and Adam who were on their honeymoon.

Sally formed her own marketing company, after gaining experience at Neal's Yard, ironically in the revamped Covent Garden. She specialized in health food companies and display work, using her natural artistic skills. Although she was completely independent of our company, she was never very far away and was always willing to lend a hand. We were very happy when she announced her engagement to Eric who, way back in 1989, had arrived from Holland as a student of horticulture. A year later, he returned to the nursery where he had become our specialist in hanging baskets.

I digress, to return to the nursery, early in the 80s, the tomato house that had been built in 1956 was showing signs of wear and tear, so between the propagating seasons, the wooden structure was dismantled and we purchased a new Robinsons aluminium propagation house. This was a slightly different shape from the original. Every square inch was valuable and had to fulfil its potential, so we invested in building some mobile benches, as seen at British Growers Look Ahead. These benches, rather than running the length of the tomato house, ran crosswise. To water or tend the crop, you moved the bench into the existing path,

thereby making a new path. By using this method we made full use of the glasshouse space. Although the house was a different shape we managed to erect the mercury phosphorus lights and when it was all completed, we stood back and thought of all those dahlias we could now grow. For sentimental reasons, I kept the old door latch – somehow aluminium houses don't have the same character as the old wooden ones.

A couple of years later my dear old bean posts, used in the construction of the Dutch light house that was situated between my original office and the packing shed, were taken down before they fell down. Robinsons came to see us, to discuss a new type of aluminium greenhouse that would be ideal for one of the crops we had in mind. This was the growing on, in spring, of specially designed

The propagation house with its mobile benching

chapter 13

hanging baskets for individual customers. This new house had a slightly different design, the gutters being further apart and the roof slightly higher and glazed with smaller sheets of glass.

Throughout the nursery's history, we had never sought the advice of any specialist garden centre designer, since David Morgan and myself were never short of ideas. But as this side of the business grew, I thought it was time to seek further advice on the future development of the retail side of the business. Having seen a very impressive presentation by an up-and-coming architect who specialized in garden centre design, we decided to take the plunge. In 1987 he came up with some very revolutionary ideas, including turning the hanging basket house that we had only completed a few years earlier into a coffee shop.

Hanging Basket House   New block of glass

He also envisaged a new sales building and a covered walkway that would join all the different departments together. The planteria was to be moved back, taking part of the dahlia field. A new block of glass was planned behind the existing greenhouses. The old propagation glasshouse would be divided into two. One half was to be the outdoor furniture showroom, while the back half was to house the new shrub advisory office, making it more accessible for customers. This was a radical revamping of the site and one that we spent many hours discussing, but we decided to go ahead and let our new advisor approach the city planning authority with his proposals. Planning permission was granted in 1988 and the preparation work was started.

The first task was to have the back block of Robinson greenhouses erected.

Needless to say the new glass had all the modern thermal screens installed and Tony and his team fitted mobile benching. The crops now grown alongside the dahlias included fuchsias, impatiens and begonias and in winter a crop of cyclamen.

Cyclamen in the glasshouse

A builder named Mr. Lelliot won the contract for the coffee shop and walkway, leaving the new retail shop to be constructed two years later. After a number of weeks Mr. Lelliot came to see me, concerned that in his opinion the walkway structure was inadequate to take the weight of the tiled roof. We took professional advice from a structural engineer colleague of Mr. Lelliot's, who confirmed the upright posts were not strong enough for the job and the whole walkway would have to be redesigned. He came up with a plan that involved strengthening the uprights with additional timbers that were bolted on. We had a very unpleasant meeting with the architect and he walked out never to be seen again. I phoned our good friend David Morgan with my tail between my legs and asked for his help; David was very understanding and took over the project.

Walkway under construction   The finished redesigned walkway

We made another huge mistake when we paved the walkway with Broadway paving. We laid them rough side up, as recommended so they were not slippery. We had never thought to question this. But we soon realized that pushing a trolley full of plants down this long path was a hazard. The vibration was so great that as well as being incredibly noisy, it also shook many of the plants out of their trays. This was a lesson we took to heart and on future extensions, the paving was placed with the other side up. Three years later we closed the walkway during the winter whilst we corrected our mistake.

The hanging basket house now had a makeover and was transformed into our new Dahlia Coffee Shop. The house was divided down the middle to accommodate a new flower shop with colour co-ordinated fittings. A brick kitchen and three

Dahlia coffee house March 1989

The flower shop

toilets were added at the far end. We were all very excited with this new venture which was opened in March 1989 and has been so hugely popular with our customers. Very soon it was realized that we would have to extend the seating area and this meant that once again the poor flower shop was on the move. It's all very well to put one's mistakes right with the benefit of hindsight, but when you are in unknown territory, it's not always easy to make the right decisions. The flower shop simply did not suit the coffee shop. When the temperature was right for one, it was wrong for the other. We were not in the least bit sorry when, two years later, the flower shop was moved back to its original updated home.

David redesigned the plan for the new sales shop and in 1991 Susan Hampshire opened it for us. Sadly, the new shop cost us our dahlia field, as the council had stipulated, in granting permission, that we should provide extra parking spaces.

Susan Hampshire at the opening of the new sales shop

the right decisions

Roger, Hazel, Sally and Julie in the dahlia field during its last season at the North Orbital Road

We had the good fortune of being able to rent two acres of agricultural land on nearby Bowmansgreen Farm, which has now been renamed Willows. The dahlias enjoy the fresh land and as it's only five minutes away from the nursery, transport is not a problem for the staff. But it certainly was an enormous decision that had to be made and one that could have been so easily solved if only I had bought more than seven and a half acres back in 1955.

... the idea of building out on piers over the ground floor

In 1993, the general office block, that I'd considered so spacious when it had first been built as a second storey building, was fast becoming overcrowded making it unworkable for our office team and stock controller. Julie and I shared a separate office, where she listened carefully (so she tells me), picking up tips on what makes our business tick and familiarizing herself with who was who in the horticultural world. Again David Morgan was called in and he came up with the idea of building out on piers, over the ground floor. He successfully gained planning permission on our behalf. This extension provided us with a brand new directors' office and boardroom, giving us more space that we could ever envisage using.

Our first grandchild, Thomas, was born in March 1993. Adam, his Father, was commuting to London daily, leaving early and often returning late in the evening. Adam was now a city trader, dealing with vast amounts of rape seed oil and a great deal of money, often travelling abroad to Argentina, or to other exotic places that took him away from home and his family.

On the rare occasions when he wasn't working while we were, such as bank holidays, he would join us for lunch. We would discuss the issues facing our business and he always listened with interest. It was at one of these lunches that he first intimated that he was tired of his jet-setting way of life and spoke of his wish to join the family business. We wondered whether he might find our life a little dull, but he assured us that he'd done quite enough travelling to last him a lifetime and was conscious that he was missing out on seeing his family. Adam joined the business in 1994, we had plenty of room for his desk in the

directors' office. He very quickly became addicted to the garden centre trade and decided that it could be just as exciting as his past career in the city. He became a director in July 1995 and was on hand when Julie presented him with a daughter, Emma, in 1996.

Every October, I would make the trip to our biggest wholesalers' trade exhibition, which was held in Sussex. It was at this show that the ordering of sundries for the coming year would take place. As an added incentive, the wholesaler ran a competition: for every order taken, a ticket was placed in a grand draw. Adam worked out that if he prepared in advance some twenty-four printed purchase orders – they were easy to generate from our stock control system – then odds were considerably in our favour. We arrived at the show early and after a hearty

Tom and Adam Wigglesworth   Emma receiving some growing tips form her Grandfather

breakfast we were ready for the off. Some eight hours later we had accomplished our mission and waited with bated breath for the draw. The system worked and we carried off numerous prizes, one of which – much to Hazel's horror – was a fifty percent ownership of a racing greyhound.

The very generous wholesaler invited us, along with the other part owner, to Wembley Stadium where we were to witness the dog running for the first time. We were very grandly entertained and all had a most enjoyable evening. This gave the staff much amusement and it was decided that when the dog ran again, we would have a table in the restaurant for any interested members. A great cheer went up when the dog won and we were all so engulfed in laughter that we were completely unaware of the loudspeaker announcement, requesting that the owner should come forward to receive the prize. I was then bundled down the hallowed Wembley steps, very conscious that receiving a prize for greyhound racing was a definite first. Our involvement with greyhound racing was not to be a long-lasting one and came to an abrupt end when the trainer was suspended for race fixing. He'd given a dog rice pudding before a race, to slow him down.

Roger Aylett with the winning Greyhound

. . . a great cheer went up when the dog won!

However, it was during a conversation with Adam that I expressed a wish to own not a greyhound but a racehorse. For the nursery's 40th anniversary, the children presented me with a part share in a wonderful horse called Broughton's Formula. This gave me a great deal of fun and the nursery staff followed the horse's career with enthusiasm. It reached its pinnacle when Formula came in at Lingfield Park, at 33 to 1. The children's idea, in giving me this present, was that it would encourage me to take time off from my busy nursery life. As I had now reached the age of sixty, they thought it was time for me to slow down a little.

The dahlia business continued to be our speciality, although gradually its financial dominance of our operation had been diminishing. Commercial dahlia growers were fast becoming an endangered species, as production costs escalated and the gardening public's tastes moved towards plants that required little or no maintenance.

Hazel and Roger Aylett with Formula at Sandown Park, in the winners enclosure

chapter 14

broughtons formula

Today, as the winter climate has become warmer, the dahlia's popularity is as strong as ever, as the tubers are often left to over-winter in the ground. We have gradually phased out exhibiting, as it created severe strains on our staffing levels back at the busy garden centre and the idea of putting up a stand of any less size or quality was unthinkable. I'm a firm believer in quitting while you are ahead and leaving a good lasting impression. I think we achieved this, but to say that I miss the thrill of winning that gold medal would be an understatement. Today, dahlias are our flagship, making our nursery and garden centre that little bit different. We still grow many thousands of plants at the trial ground at Willows Farm and during the summer

months many hundreds of bunches are sold in our garden shop. Digital photographs of each variety of dahlia that we grow have replaced the colour catalogue, enabling the customer to make a selection with an accurate picture. Modern technology used in the correct way is certainly a wonderful thing.

The dahlia greenhouse in the spring, complete with digital photographs

In September we hold an autumn garden party and the dahlias, once again are the star attraction. We fill a small marquee with their kaleidoscope of colours: it's a far cry from the old Open Days, but it demonstrates to the public that we haven't forgotten our roots.

In 1997, our electricity supply was giving us severe problems as it struggled to cope with the demand from all the different pieces of equipment. The answer was to upgrade the cable that supplied the electricity into the nursery. The size of this cable is quite amazing and I smiled to myself when I thought of that original one that had been so proudly installed, with the assurance that Marks & Spencer found it adequate and we were not in their league.

We also had at last decided to remove the old oil tank from the site. We had not needed to use oil since the conversion to gas, back in the 80's, but the tank was still full and before we could start the process of dismantling it we had to dispose of the oil. This involved hiring a specialist, to pump the oil out in a controlled manner. Having rid ourselves of this eyesore, we then addressed the problem of the boiler house site. This had been erected in the centre of the nursery, which, at the time, had made very good sense. But now it was slap bang in the middle of our very important plant retail area, taking up valuable space. To move the boiler house and its very important contents was a huge undertaking and not one that could be completed in a single season. The plan was to build an above-ground brick building to house two new boilers. This was to be sited well away from the retail area, at the back of the greenhouse that had superseded the 'guinea pig'

chapter 14

which now produces two crops a year, one of poinsettia and the other of geraniums.

Our heating engineer, Tony Abella, plus Adam and myself, spent many hours discussing the project. As stage one, the new boiler house was completed in the autumn of 1998. Then came the very necessary laying of the new gas pipe, which was to connect directly to the main supply that runs alongside the North Orbital Road.

This particular operation demonstrated to me just how far the construction industry had advanced since the first gas pipe was installed some twenty years previously. Then, it had taken many

Home grown Poinsettias are grown annually

weeks to lay, but this time I was fascinated to watch a mole machine beaver away, without even disturbing the tarmac car park which it was working under. Much to my astonishment, the whole task was completed in a single day.

The other piece of equipment that needed updating was our trusty generator. It was no longer able to supply us with sufficient electricity should the need arise, so again some re-siting seemed sensible, The generator was re-housed in a new

soundproofed building erected, much to Hazel's displeasure, just on the boundary of the bungalow's garden. Despite the sound insulation, there is no disguising the noise when this monster of a generator is running. This amazing piece of machinery has automatic switching, so when there is a power cut, apart from the noise and a slight blip on the alarm, the whole nursery and shop operation is unaffected and continues functioning in a perfectly normal manner.

Arrival of the new generator

Some two years later, in the summer of 2000, we were ready to take delivery of the new boilers. These were to be Byworth boilers, each capable of heating the whole site and having a rating of 7 million B.T.U.'s The great advantage of the new system is that if one boiler fails, the other cuts in, with no fuss or ringing of bells in the middle of the night. Well, that's how it should operate, though even modern technology lets you down occasionally. But today it's Adam that has the disturbed nights and not me!

The time had arrived when the chimney that stood in the centre of the site could be demolished. This caused quite a problem with communications around the nursery, for 'the big chimney' had been used for many years as a directional aid. Now that it was no longer there, we found ourselves having to find other

One of the new Byworth Boilers

Adam and Tom Wigglesworth surveying 'the big chimney' before its demolition landmarks when making announcements. And, sadly for them, the starlings that had nested amongst the structures for many years had to find a new home.

We had quite a tussle with the city council over the use of the area where the boiler house had stood, but in the end common sense prevailed and planning permission was granted to construct a new building as a plant information centre. This was a vast improvement on the old shed that the shrub staff had used. Incorporated into this new building was an office for our garden designer. Poor Martin, in his twenty years plus employment, had been moved from office to office and it was extremely satisfying to be able to give him a permanent home, with the correct level of light for his drawing board.

The new Plant Advisory & Information Centre   An interior view of the centre

Julie, like her Mother, manages to fit in parenting alongside a busy working life. Amongst her many tasks, which include buying and designing the Christmas displays and selecting and purchasing the garden furniture, is the important job of putting together the colour brochures for the gardening seasons. Three times a year, these are mailed to our gardening friends and serve to keep our customers informed of what is going on at Aylett Nurseries. They include details of the many different special interest events that have been arranged, along with exclusive coach trips that have been organised to some of the great gardens of England.

Thomas, Emma, Julie and Adam

These trips have been led by Brian's wife, Audrey, who deploys her special brand of professionalism to ensure that the days run smoothly. Sally has also inherited an ability to combine motherhood with a busy working life. Eric became the proud Father of twins in 2003, doubling our grandchildren with one fair stroke. Sally can often be seen with the twins in tow, as she goes about creating a display in the gift shop, or chatting to a representative about new lines that she thinks will fit in with the garden oriented gifts that she purchases.

The twins William (left) and Hannah (right), Sally and Eric

Over the years I have lost count of the number of employees that have passed through the nursery. We always say that a new member of staff will either hate the work and be gone within a very short space of time, or love it and become part of the extended 'Aylett's family'. We keep in touch with many of our past student and employees, it's always good to hear that they are successful in their careers.

Numerous members of today's staff have been with us for many years. Brian leads the way, with 49 years of loyal service; Tony has notched up 31 years; Roger Leavy, who joined us to become Shrub Manager, has completed 24 years and so has Martin Finey our Garden Designer and Paul Collins, who so admirably oversees both the geranium crop and the dahlias at Willows Farm.

I cannot mention each individual member of the 160-plus staff, but I know that without a loyal team behind me, I would never of survived, let alone Aylett Nurseries. The business is very different to the one that I first dreamt of, but creating it has given me a great deal of satisfaction. I often pinch myself, as I walk around in the evening, when the nursery is empty of both staff and customers and I remember that green field fifty years ago and all that has passed since.

And what of the future? Well, the nursery is in the safe hands of the family, but I have no crystal ball! If I'd had one, of course, life would have been much less of a challenge and far less enjoyable.

... And what of the future?

and today... 2005